Zeitschrift für Angewandte Mathematik und Mechanik

ZAMM

Applied Mathematics and Mechanics

Founded by Richard von Mises in 1921

Akademie-Verlag · Berlin

Volume 64 · 1984
Number 3

ISSN 0044-2267 ZAMM · Z. Angew. Math. u. Mech., Berlin **64** (1984) 3, 137—200 EVP 18,— M

The **aim and scope** of ZAMM is, in agreement with the intentions of its founder, to publish new results and information on applied mathematics (mainly on numerical mathematics and applications of analysis) as well as on theoretical and applied mechanics.
The journal is of interest for persons working in applied mathematics and mechanics as well as on mathematical or mechanical questions in other sciences as physics and technical sciences, for instance mechanical and civil engineering.

Terms of subscription for the journal

Orders can be sent

- in the GDR: to Postzeitungsvertrieb, or to Akademie-Verlag, DDR-1086 Berlin, Leipziger Str. 3—4, PF-Nr. 1233;
- in the other socialist countries: to a book-shop for foreign language literature or to the competent news-distributing agency;
- in the FRG and Berlin(West): to a book-shop or to the wholesale distributing agency Kunst und Wissen, Erich Bieber OHG, Wilhelmstr. 4—6, D-7000 Stuttgart 1;
- in the other Western European Countries: to Kunst und Wissen, Erich Bieber GmbH, Dufourstr. 51, CH-8008 Zürich;
- in other countries: to the international book- and journal-selling trade, to Buchexport, Volkseigener Außenhandelsbetrieb der DDR, DDR-7010 Leipzig, Postfach 160; or to the Akademie-Verlag, DDR-1086 Berlin, Leipziger Str. 3—4, PF-Nr. 1233.

Manuscripts (in grammatical English or German language, master and one copy, formulas typewritten) should be addressed to the editorial staff of ZAMM, Institut für Mechanik der Akademie der Wissenschaften der DDR, DDR-1199 Berlin, Rudower Chaussee 5. The title in English language and for papers of more than 10 pages a summary of 5 to 10 lines in English and (if possible) German and Russian language have to be added. Detailed remarks for authors which should be considered strictly are published subsequent to the last annual table of contents.

ZEITSCHRIFT FÜR
ANGEWANDTE MATHEMATIK UND MECHANIK
Herausgeber und Chefredakteur: Prof. Dr. Günter Schmidt, Berlin. Verlag: Akademie-Verlag, DDR-1086 Berlin, Leipziger Straße 3—4; Fernruf: 2236221 oder 2236229; Telex-Nr.: 114420; Bank: Staatsbank der DDR, Berlin, Kto.-Nr.: 6886-26-20712. Anschrift der Redaktion: Institut für Mechanik der Akademie der Wissenschaften der DDR, DDR-1199 Berlin, Rudower Chaussee 5; Fernruf: 6743639 oder 6743643.

Veröffentlicht unter der Lizenznummer 1282 des Presseamtes beim Vorsitzenden des Ministerrates der Deutschen Demokratischen Republik. Gesamtherstellung: VEB Druckerei „Thomas Müntzer", 5820 Bad Langensalza. Erscheinungsweise: Die Zeitschrift für Angewandte Mathematik und Mechanik erscheint monatlich. Die 12 Hefte eines Jahres einschließlich Tagungshefte bilden einen Band. Bezugspreis je Band 360,— M zuzüglich Versandspesen (Preis für die DDR 216,— M). Bezugspreis je Heft 30,— M (Preis für die DDR 18,— M). Bestellnummer dieses Heftes: 1009/64/3.

Printed in the German Democratic Republic.
AN (EDV) 35937

ZAMM 64, 137 – 146 (1984)

J. BALAŠ / V. SLÁDEK / J. SLÁDEK

The Boundary Integral Equation Method for Plates Resting on a Two-Parameter Foundation

Die Randintegralmethode wird zur Lösung eines Plattenproblems angewandt, wobei die Platte auf einem Untergrund ruht, der durch zwei Parameter beschrieben wird. Es werden drei Randwertprobleme betrachtet und drei Fälle behandelt, die sich hinsichtlich eines charakteristischen Parameters des Untergrundes unterscheiden. Die mit der Randintegralmethode erhaltene Lösung für eine eingespannte Kreisplatte mit Einzellast wird mit der analytischen Lösung verglichen.

In the paper the boundary integral equation method for solving the problem of a plate resting on a two-parameter foundation is presented. Three boundary value problems are considered, and three cases are to be distinguished as regards a characteristic parameter of the foundation. The boundary integral equation solution for a clamped circular plate under a concentrated load is compared to the analytical one.

В настоящей статье применяется метод граничных интегральных уравнений для решения задач о плитах, расположенных на основании, которое описывается при помощи двух параметров. Рассматриваются три граничные задачи и три случая, которые отличаются по характеристическому параметру основания. В работе сравниваются результаты, полученные при помощи метода граничных интегральных уравнений с результатами аналитического решения.

1. Introduction

The usual approach in formulating problems of beams, plates, and shells continuously supported by elastic media is based on the inclusion of the foundation reaction into the corresponding differential equation of the beam, plate, or shell. In attempt to find a physically close and mathematically simple representation for these materials at the constant area, one may proceed in either of two ways; namely, starting with the continuum and introducing simplifying assumptions with respect to expected displacements and/or stresses [1, 2], or starting with the Winkler foundation and, in order to bring it closer to reality, assuming some kind of interaction between the spring elements. A review of a number of foundation models as well as the proper mathematical formulation of these ones was made by KERR [3] and SELVADURAI [2].

In this paper the boundary integral equation BIE solution of the boundary value problems corresponding to a thin plate on a Pasternak-type foundation is presented. Another physical model of the two-parameter foundation, namely the Filonenko-Borodich model, is described mathematically in the same way. The Winkler model is a limiting case of both these ones [2]. In the case of a plate with a simply supported boundary or a free boundary it is necessary for the BIE to be formulated in terms of physical quantities such as the deflection, the slope on the edge, the normal bending moment and the equivalent shear force. These BIE follow from the integral representation of the deflection and its normal derivative at a point on the boundary. The latter requires a special attention as was shown by HANSEN [4] for the case of Kirchhoff's differential equation. In the case of a clamped plate boundary, however, a formulation in terms of non-physical quantities such as the Laplacian of deflection and its normal derivative is convenient. Singularity of these BIE can be handled by taking the integrals in the sense of Cauchy principle value.

Three various types of the fundamental solution of the differential equation of the problem considered are to be distinguished according to a parameter of the equation.

An example of a circular plate with the clamped boundary subjected to a concentrated load at the middle point is investigated. The analytical solution of the problem can be easily found because of the rotational symmetry of this one. The boundary integral equation method (BIEM) has also been used for evaluating deflections to compare results of this method with those of the analytical solution.

The BIEM has become increasingly popular due mainly to the wide availability of computers. As is well known the main advantage of the BIEM is the dimensional reduction of the continuum to be discretized because the unknowns are located at the boundary. The presented method is general as it is applicable to a plate with the boundary of an arbitrary shape and for arbitrary boundary conditions on any part of the boundary.

2. Fundamental Solutions

The differential equation of a thin plate of constant thickness subjected to a distributed lateral load $q(\boldsymbol{x})$ and a foundation interface pressure $p(\boldsymbol{x})$ is [3]

$$D\nabla^2\nabla^2 w = q - p \tag{1}$$

where D is the flexural rigidity of the plate and ∇^2 denotes the Laplace operator in x_1 and x_2. For the case of a plate resting on a Pasternak foundation the interface pressure is represented as

$$p = kw - G\nabla^2 w \tag{2}$$

where k and G are two parameters characterizing the foundation material.

Assuming that the plate maintains continuous contact with the base and that there are no shearing forces at the contacting surface, the differential equation of the plate assumes the form

$$\nabla^2\nabla^2 w - \frac{G}{D}\nabla^2 w + \frac{k}{D} w = \frac{q}{D}. \tag{3}$$

An additional uniform compression or tension field will affect only the constant at the Laplacian term.

Let us replace the coordinates x_α by new dimensionless variables, denoted again by x_α,

$$x_\alpha/L \to x_\alpha\,, \qquad \alpha = 1, 2$$

where L stands for a characteristic length defined as follows

$$L = (D/k)^{1/4}\,.$$

Then, the governing equation (3) becomes

$$\nabla^2\nabla^2 w - 2\varkappa\nabla^2 w + w = \frac{q}{k} \tag{4}$$

where a new dimensionless parameter $\varkappa$ is defined as

$$\varkappa = \frac{G}{2\sqrt{kD}} \geqq 0\,.$$

The differential operator of equation (4) can be split into a product of two operators of second order, namely

$$(\nabla^2 - a^2)(\nabla^2 - b^2)\, w = \frac{q}{k} \tag{5}$$

where

$$a^2 = \varkappa + (\varkappa^2 - 1)^{1/2}\,, \qquad b^2 = \varkappa - (\varkappa^2 - 1)^{1/2}\,.$$

The fundamental solution of the linear differential equation (5) is a distribution, denoted by $G(\boldsymbol{x} - \boldsymbol{y})$, which satisfies the equation

$$(\nabla^2 - a^2)(\nabla^2 - b^2)\, G(\boldsymbol{x} - \boldsymbol{y}) = \delta(\boldsymbol{x} - \boldsymbol{y})\,. \tag{6}$$

When $a^2 \neq b^2$, i.e. $\varkappa \neq 1$, the solution of equation (6) takes the form

$$G(\boldsymbol{x} - \boldsymbol{y}) = \frac{1}{a^2 - b^2}\,[g_a(\boldsymbol{x} - \boldsymbol{y}) - g_b(\boldsymbol{x} - \boldsymbol{y})] \tag{7}$$

where g_a and g_b are distributions governed by the equations

$$(\nabla^2 - a^2)\, g_a(\boldsymbol{x} - \boldsymbol{y}) = \delta(\boldsymbol{x} - \boldsymbol{y})\,, \qquad (\nabla^2 - b^2)\, g_b(\boldsymbol{x} - \boldsymbol{y}) = \delta(\boldsymbol{x} - \boldsymbol{y})\,, \tag{8}$$

respectively. Using the Fourier transform technique, one can write for $g_s(\boldsymbol{x})$, ($s = a, b$)

$$g_s(\boldsymbol{x}) = -\frac{1}{(2\pi)^2}\int \frac{e^{-i\boldsymbol{p}\cdot\boldsymbol{x}}}{p^2 + s^2}\, d^2p = -\frac{1}{2\pi}\int_0^\infty \frac{pJ_0(p|\boldsymbol{x}|)}{p^2 + s^2}\, dp = -\frac{1}{2\pi} K_0(s|\boldsymbol{x}|) \tag{9}$$

where the last integral for $\operatorname{Re} s > 0$ is taken from [5].

Thus, for $\varkappa > 1$ the fundamental solution of equation (6) is given as

$$G(\boldsymbol{x} - \boldsymbol{y}) = -\frac{1}{4\pi\sqrt{\varkappa^2 - 1}}\,[K_0(a|\boldsymbol{x} - \boldsymbol{y}|) - K_0(b|\boldsymbol{x} - \boldsymbol{y}|)]\,. \tag{10}$$

In the case of $0 \leqq \varkappa < 1$ a new parameter ψ is introduced by $\varkappa \equiv \cos 2\psi$, $\psi \in (0; \pi/4)$, and the parameter a and b are complex conjugated, $a = e^{i\psi} = \bar{b}$. Then from (7), (9) and using the relation $\overline{K_\nu(z)} = K_{\bar{\nu}}(\bar{z})$ for modified Bessel functions, one obtains

$$G(\boldsymbol{x} - \boldsymbol{y}) = -\frac{1}{2\pi \sin 2\psi} \operatorname{Im} K_0(e^{i\psi}|\boldsymbol{x} - \boldsymbol{y}|) \tag{11}$$

where $\operatorname{Im} K_0(z)$ denotes the imaginary part of $K_0(z)$. In the case of $\varkappa = 0$, which corresponds physically to a Winkler foundation, $\psi = \pi/4$ and

$$G(\boldsymbol{x} - \boldsymbol{y}) = -\frac{1}{2\pi} \operatorname{kei}(|\boldsymbol{x} - \boldsymbol{y}|)\,. \tag{12}$$

The case of $\varkappa = 1$, when $a = 1 = b$, should be examined separately. Equation (6) can be rewritten as

$$(\nabla^2 - 1)^2\, G(\boldsymbol{x} - \boldsymbol{y}) = \delta(\boldsymbol{x} - \boldsymbol{y})\,, \tag{13}$$

hence

$$G(\boldsymbol{x} - \boldsymbol{y}) = \frac{1}{(2\pi)^2}\int \frac{e^{-i\boldsymbol{p}\cdot(\boldsymbol{x}-\boldsymbol{y})}}{(p^2 + 1)^2}\, d^2p = \frac{1}{2\pi}\int_0^\infty \frac{pJ_0(p|\boldsymbol{x} - \boldsymbol{y}|)}{(p^2 + 1)^2}\, dp = \frac{1}{4\pi}|\boldsymbol{x} - \boldsymbol{y}|\, K_1(|\boldsymbol{x} - \boldsymbol{y}|) \tag{14}$$

where the last integral can be found in [5].

The fundamental solutions given by equations (10), (11) and (14) are necessary for an integral expression of the deflection governed by equation (4).

3. Boundary Integral Equation Formulation

It is convenient to rewrite equation (4) into the form

$$K_{\alpha\beta\gamma\delta}\, w_{,\alpha\beta\gamma\delta} - 2\varkappa\nabla^2 w + w = \frac{q}{k} \tag{15}$$

where

$$K_{\alpha\beta\gamma\delta} \equiv (1-\nu)\,\delta_{\alpha\gamma}\delta_{\beta\delta} + \nu\delta_{\alpha\beta}\delta_{\gamma\delta}\,, \tag{16}$$

ν is Poisson's ratio, and subscripts following the comma denote the partial differentiation with respect to associated coordinate, and repeated Greek subscripts are to be summed over the range (1, 2).

According to equation (15) one can write

$$\int\limits_S G(\boldsymbol{x}-\boldsymbol{y})\,\frac{q(\boldsymbol{y})}{k}\,\mathrm{d}S_y = \int\limits_S G(\boldsymbol{x}-\boldsymbol{y})\,[K_{\alpha\beta\gamma\delta}\, w_{,\alpha\beta\gamma\delta} - 2\varkappa\nabla^2 w + w]\,\mathrm{d}S_y\,.$$

Having used the two-dimensional form of the Gauss' divergence theorem several times, the last equation can be rewritten in terms of contour integrals

$$\begin{aligned}\int\limits_S G(\boldsymbol{x}-\boldsymbol{y})\,\frac{q(\boldsymbol{y})}{k}\,\mathrm{d}S_y = \int\limits_C & K_{\alpha\beta\gamma\delta}[G(\boldsymbol{x}-\boldsymbol{\eta})\, w_{,\beta\gamma\delta}(\boldsymbol{\eta})\, n_\alpha(\boldsymbol{\eta}) - G_{,\alpha}(\boldsymbol{x}-\boldsymbol{\eta})\, w_{,\gamma\delta}(\boldsymbol{\eta})\, n_\beta(\boldsymbol{\eta}) + \\ & + G_{,\alpha\beta}(\boldsymbol{x}-\boldsymbol{\eta})\, w_{,\delta}(\boldsymbol{\eta})\, n_\gamma(\boldsymbol{\eta}) - G_{,\alpha\beta\gamma}(\boldsymbol{x}-\boldsymbol{\eta})\, n_\delta(\boldsymbol{\eta})\, w(\boldsymbol{\eta})]\,\mathrm{d}l_\eta - \\ & - 2\varkappa\int\limits_C \left[G(\boldsymbol{x}-\boldsymbol{\eta})\,\frac{\partial w(\boldsymbol{\eta})}{\partial \boldsymbol{n}(\boldsymbol{\eta})} - w(\boldsymbol{\eta})\,\frac{\partial G(\boldsymbol{x}-\boldsymbol{\eta})}{\partial \boldsymbol{n}(\boldsymbol{\eta})}\right]\mathrm{d}l_\eta + \\ & + \int\limits_S w(\boldsymbol{y})[\nabla^2\nabla^2 - 2\varkappa\nabla^2 + 1]\, G(\boldsymbol{x}-\boldsymbol{y})\,\mathrm{d}S_y\end{aligned} \tag{17}$$

where $n_\alpha(\boldsymbol{\eta})$ are components of the outward unit normal on C, and $\boldsymbol{\eta}$ denotes a point on the boundary C. According to equation (6) the last integral in (17) equals the deflection $w(\boldsymbol{x})$, i.e.

$$\int\limits_S w(\boldsymbol{y})[\nabla^2\nabla^2 - 2\varkappa\nabla^2 + 1]\, G(\boldsymbol{x}-\boldsymbol{y})\,\mathrm{d}S_y = w(\boldsymbol{x})\,. \tag{18}$$

Now we introduce the following definitions

$$\left.\begin{aligned} & M_{\alpha\beta}(\boldsymbol{x}) \equiv -\frac{D}{L^2} K_{\alpha\beta\gamma\delta}\, w_{,\gamma\delta}(\boldsymbol{x}) \equiv \frac{D}{L^2}\tilde{M}_{\alpha\beta}(\boldsymbol{x})\,, \qquad N_\alpha(\boldsymbol{x}) \equiv -\frac{D}{L^3} K_{\alpha\beta\gamma\delta}\, w_{,\beta\gamma\delta}(\boldsymbol{x}) \equiv \frac{D}{L^3}\tilde{N}_\alpha(\boldsymbol{x})\,, \\ & \mathcal{M}_{\alpha\beta}(\boldsymbol{x}-\boldsymbol{y}) \equiv -\frac{D}{L^2} K_{\alpha\beta\gamma\delta}\, G_{,\gamma\delta}(\boldsymbol{x}-\boldsymbol{y}) \equiv \frac{D}{L^2}\tilde{\mathcal{M}}_{\alpha\beta}(\boldsymbol{x}-\boldsymbol{y})\,, \\ & \mathcal{N}_\alpha(\boldsymbol{x}-\boldsymbol{y}) \equiv -\frac{D}{L^3} K_{\alpha\beta\gamma\delta}\, G_{,\beta\gamma\delta}(\boldsymbol{x}-\boldsymbol{y}) \equiv \frac{D}{L^3}\tilde{\mathcal{N}}_\alpha(\boldsymbol{x}-\boldsymbol{y})\,, \\ & M_{nn} \equiv M_{\alpha\beta}n_\alpha n_\beta\,, \qquad M_{nt} \equiv M_{\alpha\beta}n_\alpha t_\beta\,, \qquad H_n \equiv N_n + \frac{\partial}{\partial t} M_{nt}\,. \end{aligned}\right\} \tag{19}$$

$\mathcal{M}_{nn}$, $\mathcal{M}_{nt}$, $\mathcal{H}_n$ are defined in full analogy, and $\partial/\partial \boldsymbol{t}$ denotes $t_\alpha(\boldsymbol{\eta})\,\partial/\partial\eta_\alpha$. Some of these quantities have a physical meaning:

$$M_{nn} = -\frac{D}{L^2}[(1-\nu)\, w_{,\alpha\beta}n_\alpha n_\beta + \nu\nabla^2 w]$$

is the normal bending moment,

$$M_{nt} = -\frac{D}{L^2}(1-\nu)\, w_{,\alpha\beta}n_\alpha t_\beta$$

is the twisting moment,

$$N_n = -\frac{D}{L^3} n_\alpha \nabla^2 w_{,\alpha}$$

is the transverse shear force, and H_n is the Kirchhoff equivalent shear force.

Since the normal and unit tangent at a point on C make complete orthogonal system of vectors, we have

$$n_\alpha n_\beta + t_\alpha t_\beta = \delta_{\alpha\beta}\,. \tag{20}$$

Using (20) and (19), the following rearrangements can be performed:

$$\int_C K_{\alpha\beta\gamma\delta} G_{,\alpha} w_{,\gamma\delta} n_\beta \, dl_\eta = -\int_C \left[\tilde{M}_{nn} \frac{\partial G}{\partial n} + \tilde{M}_{nt} \frac{\partial G}{\partial t}\right] dl_\eta , \tag{21}$$

$$\int_C K_{\alpha\beta\gamma\delta} G_{,\alpha\beta} w_{,\delta} n_\gamma \, dl_\eta = -\int_C \left[\tilde{\mathcal{M}}_{nn} \frac{\partial w}{\partial n} + \tilde{\mathcal{M}}_{nt} \frac{\partial w}{\partial t}\right] dl_\eta . \tag{22}$$

Let the boundary C have a continuous tangent, then

$$\left.\begin{aligned} \int_C \tilde{M}_{nt}(\eta) \frac{\partial G(x - \eta)}{\partial t(\eta)} \, dl_\eta &= -\int_C \frac{\partial \tilde{M}_{nt}(\eta)}{\partial t(\eta)} G(x - \eta) \, dl_\eta , \\ \int_C \tilde{\mathcal{M}}_{nt}(\eta) \frac{\partial w(\eta)}{\partial t(\eta)} \, dl_\eta &= -\int_C \frac{\partial \tilde{\mathcal{M}}_{nt}(x - \eta)}{\partial t(\eta)} w(\eta) \, dl_\eta . \end{aligned}\right\} \tag{23}$$

Finally, from (17)—(23) one obtains the integral representation of the deflection

$$w(x) = \int_S G(x - y) \frac{q(y)}{k} \, dS_y + \int_C \left\{\tilde{H}_n(\eta) G(x - \eta) - \tilde{M}_{nn}(\eta) \frac{\partial G(x - \eta)}{\partial n(\eta)} - \right.$$

$$\left. - w(\eta) \left[\tilde{\mathcal{H}}_n(x - \eta) + 2\varkappa \frac{\partial G(x - \eta)}{\partial n(\eta)}\right] + \frac{\partial w(\eta)}{\partial n(\eta)} [\tilde{\mathcal{M}}_{nn}(x - \eta) + 2\varkappa G(x - \eta)]\right\} dl_\eta . \tag{24}$$

There is another integral representation of the deflection given by equation (40), besides this general one. One of them is chosen according to boundary conditions. Only some of the boundary values of the quantities $\tilde{H}_n(\eta)$, $\tilde{M}_{nn}(\eta)$, $w(\eta)$ and $\partial w/\partial n$ are prescribed as boundary conditions.

It can be shown for any $\varkappa \geq 0$ that all potentials in equation (24) behave continuously, when $S \ni x \to \xi \in C$, except those with the kernel $\partial K_0(|x - \eta|)/\partial n(\eta)$, the behaviour of which is the same as that of potentials with the kernel $\partial \ln |x - \eta|/\partial n(\eta)$. Thus from equation (24) the first BIE follows

$$\frac{1}{2} w(\xi) = \int_S G(\xi - y) \frac{q(y)}{k} \, dS_y + \int_C \left\{\tilde{H}_n(\eta) G(\xi - \eta) - \tilde{M}_{nn}(\eta) \frac{\partial G(\xi - \eta)}{\partial n(\eta)} - \right.$$

$$\left. - w(\eta) \left[\tilde{\mathcal{H}}_n(\xi - \eta) + 2\varkappa \frac{\partial G(\xi - \eta)}{\partial n(\eta)}\right] + \frac{\partial w(\eta)}{\partial n(\eta)} [\tilde{\mathcal{M}}_{nn}(\xi - \eta) + 2\varkappa G(\xi - \eta)]\right\} dl_\eta \tag{25}$$

where $w(x)$ is satisfying the Hölder condition and boundary C is smooth. All the integrals exist in the sense of Cauchy principal value at least.

a) Simply supported boundary

For the case of a simply supported edge the boundary conditions $w(\eta) = 0$ and $M_{nn}(\eta) = 0$ together with equation (24) give the integral representation of the deflection

$$w(x) = \int_S G(x - y) \frac{q(y)}{k} \, dS_y + \int_C \left\{\tilde{H}_n(\eta) G(x - \eta) + \frac{\partial w(\eta)}{\partial n(\eta)} [\tilde{\mathcal{M}}_{nn}(x - \eta) + 2\varkappa G(x - \eta)\right] dl_\eta . \tag{26}$$

Two BIE are necessary to determine the unknown boundary values $\tilde{H}_n(\eta)$ and $\partial w(\eta)/\partial n(\eta)$. The first one follows directly from equation (25)

$$0 = \int_S G(\xi - y) \frac{q(y)}{k} \, dS_y + \int_C \left\{\tilde{H}_n(\eta) G(\xi - \eta) + \frac{\partial w(\eta)}{\partial n(\eta)} [\tilde{\mathcal{M}}_{nn}(\xi - \eta) + 2\varkappa G(\xi - \eta)]\right\} dl_\eta . \tag{27}$$

The second BIE can be obtained from the directional derivative of equation (26) by carrying out the limit procedure $S \ni x \to \xi \in C$. The differentiation of all the integrals can be carried out behind the integral signs even if $x \to \xi$, as the integrals exist in the sense of Cauchy principal value. Considering the limit behaviour of kernels, one can find the second BIE

$$\frac{1}{2} \frac{\partial w(\xi)}{\partial n(\xi)} = \int_S \frac{\partial G(\xi - y)}{\partial n(\xi)} \frac{q(y)}{k} \, dS_y + \int_C \left\{\tilde{H}_n(\eta) \frac{\partial G(\xi - \eta)}{\partial n(\xi)} + \right.$$

$$\left. - \frac{\partial w(\eta)}{\partial n(\eta)} \frac{\partial}{\partial n(\xi)} [\tilde{\mathcal{M}}_{nn}(\xi - \eta) + 2\varkappa G(\xi - \eta)]\right\} dl_\eta . \tag{28}$$

The kernels $\partial G/\partial \boldsymbol{n}$ for several $\varkappa$ are given by

$$\frac{\partial G(\xi-\eta)}{\partial n(\xi)} = \begin{cases} \frac{1}{4\pi\sqrt{\varkappa^2-1}} \frac{\xi_\alpha - \eta_\alpha}{|\xi-\eta|} n_\alpha(\xi)\,[aK_1(a|\xi-\eta|) - bK_1(b|\xi-\eta|)]\,, & \text{for} \quad \varkappa > 1\,, \\ \frac{1}{2\pi \sin 2\psi} \frac{\xi_\alpha - \eta_\alpha}{|\xi-\eta|} n_\alpha(\xi)\, \mathrm{Im}\,[\mathrm{e}^{i\psi} K_1(\mathrm{e}^{i\psi}|\xi-\eta|)]\,, & \text{for} \quad 0 \leqq \varkappa < 1\,, \\ -\frac{1}{4\pi}(\xi_\alpha - \eta_\alpha) n_\alpha(\xi)\, K_0(|\xi-\eta|)\,, & \text{for} \quad \varkappa = 1\,. \end{cases} \tag{29}$$

Having solved the BIE (27) and (28) for unknown $\widetilde{H}_n(\eta)$ and $\partial w(\eta)/\partial n(\eta)$, equation (26) can be used to calculate the deflection of a plate resting on a continuous foundation with a simply supported edge of an arbitrary shape.

b) Free boundary

In this case there are two regions which interact at the boundary C. The plate region S with deflections w_p governed by equation (4) and the surrounding foundation region $\overline{S}$ with deflections w_F described by the second order equation (2), which can be written in dimensionless variables x_α as

$$\nabla^2 w_F(\boldsymbol{x}) - \frac{1}{2\varkappa} w_F(\boldsymbol{x}) = 0\,, \quad \text{for} \quad \boldsymbol{x} \in \overline{S}\,. \tag{30}$$

As was shown by KERR [3] the necessary three boundary conditions for edge may be written as follows

$$w_p(\eta) = w_F(\eta)\,, \qquad M_{nn}(\eta) = 0\,, \qquad H_n(\eta) + G\frac{\partial w_p(\eta)}{\partial n(\eta)} = G\frac{\partial w_F(\eta)}{\partial n(\eta)}. \tag{31}$$

The integral expression of $w_F(\boldsymbol{x})$ for $\boldsymbol{x} \in \overline{S}$ can be immediately written as

$$w_F(\boldsymbol{x}) = \frac{1}{2\pi}\int_C \left[\frac{\partial w_F(\eta)}{\partial n(\eta)} K_0(\gamma|\boldsymbol{x}-\eta|) - w_F(\eta)\frac{\partial K_0(\gamma|\boldsymbol{x}-\eta|)}{\partial n(\eta)}\right] \mathrm{d}l_\eta \tag{32}$$

where

$$\gamma = (1/2\varkappa)^{1/2}\,.$$

Solutions of equations (4) and (30) with the boundary conditions (31) are expressed in an integral representation as follows

$$w_p(\boldsymbol{x}) = \int_S G(\boldsymbol{x}-\boldsymbol{y})\frac{q(\boldsymbol{y})}{k}\,\mathrm{d}S_y + \int_C \Big\{\widetilde{H}_n(\eta)\, G(\boldsymbol{x}-\eta) - w_p(\eta)\Big[\widetilde{\mathscr{H}}_n(\boldsymbol{x}-\eta) + {}$$

$$+ 2\varkappa\frac{\partial G(\boldsymbol{x}-\eta)}{\partial n(\eta)}\Big] + \frac{\partial w_p(\eta)}{\partial n(\eta)}[\widetilde{\mathscr{M}}_{nn}(\boldsymbol{x}-\eta) + 2\varkappa G(\boldsymbol{x}-\eta)]\Big\}\,\mathrm{d}l_\eta\,, \qquad \boldsymbol{x} \in S\,, \tag{33}$$

$$w_F(\boldsymbol{x}) = \frac{1}{2\pi}\int_C \left\{\left[\frac{D}{GL^3}\widetilde{H}_n(\eta) + \frac{\partial w_p(\eta)}{\partial n(\eta)}\right] K_0(\gamma|\boldsymbol{x}-\eta|) - w_p(\eta)\frac{\partial K_0(\gamma|\boldsymbol{x}-\eta|)}{\partial n(\eta)}\right\}\mathrm{d}l_\eta\,, \qquad \boldsymbol{x} \in \overline{S}\,. \tag{34}$$

Approaching $\overline{S} \ni \boldsymbol{x} \to \xi \in C$ in equation (34), one finds the BIE

$$\frac{1}{2} w_p(\xi) = \frac{1}{2\pi}\int_C \left\{\left[\frac{D}{GL^3}\widetilde{H}_n(\eta) + \frac{\partial w_p(\eta)}{\partial n(\eta)}\right] K_0(\gamma|\xi-\eta|) - w_p(\eta)\frac{\partial K_0(\gamma|\xi-\eta|)}{\partial n(\eta)}\right\}\mathrm{d}l_\eta\,. \tag{35}$$

Another BIE follows from equations (25) and (31)

$$\frac{1}{2} w_p(\xi) = \int_S G(\xi-\boldsymbol{y})\frac{q(\boldsymbol{y})}{k}\,\mathrm{d}S_y + \int_C \Big\{\widetilde{H}_n(\eta)\, G(\xi-\eta) - w_p(\eta)\Big[\widetilde{\mathscr{H}}_n(\xi-\eta) + {}$$

$$+ 2\varkappa\frac{\partial G(\xi-\eta)}{\partial n(\eta)}\Big] + \frac{\partial w_p(\eta)}{\partial n(\eta)}[\widetilde{\mathscr{M}}_{nn}(\xi-\eta) + 2\varkappa G(\xi-\eta)]\Big\}\,\mathrm{d}l_\eta\,. \tag{36}$$

The last BIE can be obtained from the directional derivative of equation (34). The differentiation of the last integral in (34) cannot be carried out behind the integral sign when $\boldsymbol{x} \to \xi$ because such differentiated kernel behaves like r^{-2}. Thus the third BIE takes the form

$$\frac{1}{2}\left[\frac{D}{GL^3}\widetilde{H}_n(\xi) + \frac{\partial w_p(\xi)}{\partial n(\xi)}\right] = \frac{1}{2\pi}\int_C \left[\frac{D}{GL^3}\widetilde{H}_n(\eta) + \frac{\partial w_p(\eta)}{\partial n(\eta)}\right]\frac{\partial K_0(\gamma|\xi-\eta|)}{\partial n(\xi)}\,\mathrm{d}l_\eta - {}$$

$$- \frac{n_\alpha(\xi)}{2\pi}\lim_{\boldsymbol{x}\to\xi}\frac{\partial}{\partial x_\alpha}\int_C w_p(\eta)\frac{\partial K_0(\gamma|\boldsymbol{x}-\eta|)}{\partial n(\eta)}\,\mathrm{d}l_\eta\,. \tag{37}$$

The last term in equation (37) can be handled in the same manner as was proposed for a similar one by HANSEN [4].

The three unknown quantities $w_p(\eta)$, $\partial w_p/\partial n(\eta)$ and $\tilde{H}_n(\eta)$ have to be evaluated from equations (35)—(37) before using the relations (33) and (34).

c) Clamped boundary

In the case of a clamped boundary another integral representation of the deflection in terms of w, $\partial w/\partial \boldsymbol{n}$, $\nabla^2 w$ and $\partial\nabla^2 w/\partial \boldsymbol{n}$ on the boundary C is found to be appropriate. For this purpose the following simple relations are written

$$\left.\begin{aligned} &GK_{\alpha\beta\gamma\delta}\, w_{,\beta\gamma\delta} n_\alpha = G\frac{\partial\nabla^2 w}{\partial \boldsymbol{n}}, \\ &K_{\alpha\beta\gamma\delta}\, G_{,\alpha}\, w_{,\gamma\delta} n_\beta = \nabla^2 w\frac{\partial G}{\partial \boldsymbol{n}} + (1-\nu)\left[G_{,\alpha} w_{,\alpha\beta} n_\beta - \nabla^2 w\frac{\partial G}{\partial \boldsymbol{n}}\right], \\ &K_{\alpha\beta\gamma\delta}\, G_{,\alpha\beta}\, w_{,\delta} n_\gamma = \nabla^2 G\frac{\partial w}{\partial \boldsymbol{n}} + (1-\nu)\left[n_\alpha\, G_{,\alpha\beta} w_{,\beta} - \nabla^2 G\frac{\partial w}{\partial \boldsymbol{n}}\right], \\ &K_{\alpha\beta\gamma\delta}\, G_{,\alpha\beta\gamma}\, w n_\delta = w\frac{\partial\nabla^2 G}{\partial \boldsymbol{n}}. \end{aligned}\right\} \tag{38}$$

The auxiliary identity

$$\int_C [G_{,\alpha} w_{,\alpha\beta} n_\beta - n_\alpha G_{,\alpha} w_{,\beta\beta} - n_\alpha G_{,\alpha\beta} w_{,\beta} + G_{,\alpha\alpha} w_{,\beta} n_\beta]\, \mathrm{d}l_\eta = 0 \tag{39}$$

is a direct consequence of the Gauss' divergence theorem. From (17), (18), (38) and (39) one obtains the following integral expression of the deflection

$$\begin{aligned} w(\boldsymbol{x}) = \int_S G(\boldsymbol{x}-\boldsymbol{y})\frac{q(\boldsymbol{y})}{k}\,\mathrm{d}S_y + \int_C \Big\{ &-\frac{\partial\nabla^2 w(\boldsymbol{\eta})}{\partial \boldsymbol{n}(\boldsymbol{\eta})} G(\boldsymbol{x}-\boldsymbol{\eta}) + \nabla^2 w(\boldsymbol{\eta})\frac{\partial G(\boldsymbol{x}-\boldsymbol{\eta})}{\partial \boldsymbol{n}(\boldsymbol{\eta})} - \\ &- \frac{\partial w(\boldsymbol{\eta})}{\partial \boldsymbol{n}(\boldsymbol{\eta})}[\nabla^2 G(\boldsymbol{x}-\boldsymbol{\eta}) - 2\varkappa G(\boldsymbol{x}-\boldsymbol{\eta})] + w(\boldsymbol{\eta})\left[\frac{\partial\nabla^2 G(\boldsymbol{x}-\boldsymbol{\eta})}{\partial \boldsymbol{n}(\boldsymbol{\eta})} - 2\varkappa\frac{\partial G(\boldsymbol{x}-\boldsymbol{\eta})}{\partial \boldsymbol{n}(\boldsymbol{\eta})}\right]\Big\}\,\mathrm{d}l_\eta. \end{aligned} \tag{40}$$

Taking into account the boundary conditions $w(\boldsymbol{\eta}) = 0$ and $\partial w/\partial \boldsymbol{n}(\boldsymbol{\eta}) = 0$, the last equation can be rewritten as

$$w(\boldsymbol{x}) = \int_S G(\boldsymbol{x}-\boldsymbol{y})\frac{q(\boldsymbol{y})}{k}\,\mathrm{d}S_y + \int_C \left[\nabla^2 w(\boldsymbol{\eta})\frac{\partial G(\boldsymbol{x}-\boldsymbol{\eta})}{\partial \boldsymbol{n}(\boldsymbol{\eta})} - \frac{\partial\nabla^2 w(\boldsymbol{\eta})}{\partial \boldsymbol{n}(\boldsymbol{\eta})} G(\boldsymbol{x}-\boldsymbol{\eta})\right]\mathrm{d}l_\eta. \tag{41}$$

The non-physical quantities $\nabla^2 w$ and $\partial\nabla^2 w/\partial \boldsymbol{n}$ on the boundary C are to be determined before applying equation (41). For this purpose let us apply the Laplace operator to both sides of equation (41) with the result

$$\nabla^2 w(\boldsymbol{x}) = \int_S \nabla^2 G(\boldsymbol{x}-\boldsymbol{y})\frac{q(\boldsymbol{y})}{k}\,\mathrm{d}S_y + \int_C \left[\nabla^2 w(\boldsymbol{\eta})\frac{\partial\nabla^2 G(\boldsymbol{x}-\boldsymbol{\eta})}{\partial \boldsymbol{n}(\boldsymbol{\eta})} - \frac{\partial\nabla^2 w(\boldsymbol{\eta})}{\partial \boldsymbol{n}(\boldsymbol{\eta})}\nabla^2 G(\boldsymbol{x}-\boldsymbol{\eta})\right]\mathrm{d}l_\eta \tag{42}$$

where

$$\nabla^2 G(\boldsymbol{x}-\boldsymbol{y}) = \begin{cases} \dfrac{1}{2\pi(a^2-b^2)}\,[b^2 K_0(b|\boldsymbol{x}-\boldsymbol{y}|) - a^2 K_0(a|\boldsymbol{x}-\boldsymbol{y}|)], & \text{for } \varkappa \neq 1 \\ \dfrac{1}{4\pi}|\boldsymbol{x}-\boldsymbol{y}|\,K_1(|\boldsymbol{x}-\boldsymbol{y}|) - \dfrac{1}{2\pi}K_0(|\boldsymbol{x}-\boldsymbol{y}|), & \text{for } \varkappa = 1 \end{cases} \tag{43}$$

Considering the limit behaviour of potentials in (41) and (42) and the boundary conditions $w(\boldsymbol{\eta}) = 0$, one obtains the following BIE

$$0 = \int_S G(\boldsymbol{\xi}-\boldsymbol{y})\frac{q(\boldsymbol{y})}{k}\,\mathrm{d}S_y + \int_C \left[\nabla^2 w(\boldsymbol{\eta})\frac{\partial G(\boldsymbol{\xi}-\boldsymbol{\eta})}{\partial \boldsymbol{n}(\boldsymbol{\eta})} - \frac{\partial\nabla^2 w(\boldsymbol{\eta})}{\partial \boldsymbol{n}(\boldsymbol{\eta})} G(\boldsymbol{\xi}-\boldsymbol{\eta})\right]\mathrm{d}l_\eta, \tag{44}$$

$$\frac{\alpha(\boldsymbol{\xi})}{2\pi}\nabla^2 w(\boldsymbol{\xi}) = \int_S \nabla^2 G(\boldsymbol{\xi}-\boldsymbol{y})\frac{q(\boldsymbol{y})}{k}\,\mathrm{d}S_y + \int_C \left[\nabla^2 w(\boldsymbol{\eta})\frac{\partial\nabla^2 G(\boldsymbol{\xi}-\boldsymbol{\eta})}{\partial \boldsymbol{n}(\boldsymbol{\eta})} - \frac{\partial\nabla^2 w(\boldsymbol{\eta})}{\partial \boldsymbol{n}(\boldsymbol{\eta})}\nabla^2 G(\boldsymbol{\xi}-\boldsymbol{\eta})\right]\mathrm{d}l_\eta. \tag{45}$$

In this formulation the notion of corners on the boundary C is also permitted. There are no troubles with non-unique definition of normal at a corner. The quantity $\alpha(\boldsymbol{\xi})$ denotes the angle between tangents at $\boldsymbol{\xi} \in C$. If the boundary contour C is smooth at $\boldsymbol{\xi}$, $\alpha(\boldsymbol{\xi}) = \pi$.

These BIE serve to evaluate the Laplacian of the deflection and its normal derivative required in (41). The contour integrals in (44) and (45) exist in the sense of Cauchy principal value at least, and they do not require such special attention as equation (37).

In the case of mixed boundary conditions the integral representation (24) must be used. The not-prescribed quantities on several parts of the boundary C are to be calculated by a system of BIE. There is no difficulty to construct such a system by using (25) and methods described in the parts a) and b).

4. Circular Plate

a) Analytical solution

A circular plate under a rotationally symmetrical load and with the boundary conditions of the same symmetry represents a rotationally symmetrical problem. Consequently all physical quantities are only functions of the radius r in polar coordinates. The problem is reduced to a one-dimensional one, and the partial differential equation (4) to the ordinary one with the Laplace operator given by

$$\nabla^2 = \frac{\mathrm{d}^2}{\mathrm{d}r^2} + \frac{1}{r}\frac{\mathrm{d}}{\mathrm{d}r}.$$

Then the total solution of equation (4) consists of a particular solution and the solution of the homogeneous differential equation

$$w(r) = w_{\text{part}}(r) + w_{\text{hom}}(r)\,. \tag{46}$$

Up to now we have used the Cartesian coordinate system. In a curvilinear coordinate system the contravariant components of the moment tensor and shearing force are expressed as

$$M^{\alpha\beta}(\boldsymbol{x}) = -\frac{D}{L^2} K^{\alpha\beta\gamma\delta} w|_{\gamma\delta}\,, \qquad N^{\alpha}(\boldsymbol{x}) = -\frac{D}{L^3} K^{\alpha\beta\gamma\delta} w|_{\beta\gamma\delta} \tag{47}$$

with

$$K^{\alpha\beta\gamma\delta} = (1-\nu)\, g^{\alpha\gamma} g^{\beta\delta} + \nu g^{\alpha\beta} g^{\gamma\delta}$$

where $g^{\alpha\beta}$ are contravariant components of the metric tensor, and $w|_{\alpha\beta}$ denotes a covariant derivative [6]. Considering the rotational symmetry, one finds for physical components of these quantities in polar coordinate system

$$\left.\begin{aligned} &M^{rr}(r) = -\frac{D}{L^2}\left(\nabla^2 w - \frac{1-\nu}{r} w_{,r}\right), \qquad M^{\varphi\varphi}(r) = -\frac{D}{L^2}\left(\nu\nabla^2 w + \frac{1-\nu}{r} w_{,r}\right),\\ &M^{r\varphi}(r) = 0 = N^{\varphi}(r)\,, \qquad N^{r}(r) = -\frac{D}{L^3}\frac{\mathrm{d}}{\mathrm{d}r}\nabla^2 w\,. \end{aligned}\right\} \tag{48}$$

Since the normal and tangent on the boundary of a circular plate have components $n_r = 1$ and $n_\varphi = 0$, $t_r = 0$ and $t_\varphi = 1$,

$$\left.\begin{aligned} &M_{nn}(R) = M^{\alpha\beta}(R)\, n_\alpha n_\beta = M^{rr}(R)\,, \qquad M_{nt}(R) = M^{\alpha\beta}(R)\, n_\alpha t_\beta = M^{r\varphi}(R) = 0\,,\\ &N_n(R) = N^{\alpha}(R)\, n_\alpha = N^{r}(R)\,. \end{aligned}\right\} \tag{49}$$

According to equation (46) and the linearity of (48), moments and shearing forces consist of two parts

$$M^{\alpha\beta}(r) = M^{\alpha\beta}_{\text{part}}(r) + M^{\alpha\beta}_{\text{hom}}(r)\,, \qquad N^{r}(r) = N^{r}_{\text{part}}(r) + N^{r}_{\text{hom}}(r)\,. \tag{50}$$

In what follows we shall consider the concentrated load at the middle point $r = 0$, i.e. $q(\boldsymbol{r}) = q\delta(\boldsymbol{r})$. Then the particular solutions for several $\varkappa$ are given by corresponding fundamental solutions found in Sec. 2. The non-singular solutions of the homogeneous equations in the region $r \leqq R$ are solutions of Bessel equations. Thus the deflections (46) take the form

$$w(r) = A_1 I_0(ar) + A_2 I_0(br) - \frac{q}{4\pi k\sqrt{\varkappa^2-1}}\left[K_0(ar) - K_0(br)\right], \quad \text{for } \varkappa > 1\,, \tag{51}$$

$$w(r) = A_1 I_0(r) + A_2 r I_1(r) + \frac{q}{4\pi k} r K_1(r)\,, \quad \text{for} \quad \varkappa = 1\,, \tag{52}$$

$$w(r) = A_1 I_{0r}(r\,\mathrm{e}^{i\psi}) + A_2 I_{0i}(r\,\mathrm{e}^{i\psi}) - \frac{q}{2\pi k \sin 2\psi} K_{0i}(r\,\mathrm{e}^{i\psi})\,, \quad \text{for} \quad 0 \leqq \varkappa < 1\,, \tag{53}$$

where subscripts r and i denote the real and imaginary part of a quantity respectively.

The bending moments and shearing forces corresponding to the deflections given by (51)—(53) are

$$\begin{aligned} M^{rr}(r) = &-\frac{D}{L^2}\left\{A_1\left[a^2 I_0(ar) - \frac{1-\nu}{r} a I_1(ar)\right] + A_2\left[b^2 I_0(br) - \frac{1-\nu}{r} b I_1(br)\right]\right\} +\\ &+ \frac{qD}{4\pi k L^2\sqrt{\varkappa^2-1}}\left\{a^2 K_0(ar) + \frac{1-\nu}{r} a K_1(ar) - b^2 K_0(br) - \frac{1-\nu}{r} b K_1(br)\right\}, \end{aligned} \tag{54}$$

$$\begin{aligned} M^{\varphi\varphi}(r) = &-\frac{D}{L^2}\left\{A_1\left[\nu a^2 I_0(ar) + \frac{1-\nu}{r} a I_1(ar)\right] + A_2\left[\nu b^2 I_0(br) + \frac{1-\nu}{r} b I_1(br)\right]\right\} +\\ &+ \frac{qD}{4\pi k L^2\sqrt{\varkappa^2-1}}\left\{\nu a^2 K_0(ar) - \frac{1-\nu}{r} a K_1(ar) - \nu b^2 K_0(br) + \frac{1-\nu}{r} b K_1(br)\right\}, \end{aligned} \tag{55}$$

$$N^{r}(r) = -\frac{D}{L^3}\left\{A_1 a^3 I_1(ar) + A_2 b^3 I_1(br) + \frac{q}{4\pi k\sqrt{\varkappa^2-1}}\left[a^3 K_1(ar) - b^3 K_1(br)\right]\right\} \tag{56}$$

for $\varkappa > 1$,

$$M^{rr}(r) = -\frac{D}{L^2}\left\{A_1\left[I_0(r) - \frac{1-\nu}{r}I_1(r)\right] + A_2[(1+\nu)\,I_0(r) + rI_1(r)] + \right.$$

$$\left. + \frac{q}{4\pi k}[-(1+\nu)\,K_0(r) + rK_1(r)]\right\}, \tag{57}$$

$$M^{\varphi\varphi}(r) = -\frac{D}{L^2}\left\{A_1\left[\nu I_0(r) + \frac{1-\nu}{r}I_1(r)\right] + A_2[(1+\nu)\,I_0(r) + \nu rI_1(r)] + \right.$$

$$\left. + \frac{q}{4\pi k}[-(1+\nu)\,K_0(r) + \nu rK_1(r)]\right\}, \tag{58}$$

$$N^r(r) = -\frac{D}{L^3}\left\{A_1I_1(r) + A_2[2I_1(r) + rI_0(r)] + \frac{q}{4\pi k}[2K_1(r) - rK_0(r)]\right\} \tag{59}$$

for $\varkappa = 1$, and

$$M^{rr}(r) = -\frac{D}{L^2}\left\{A_1\left[\cos 2\psi I_{0r}(z) - \sin 2\psi I_{0i}(z) - \frac{1-\nu}{r}(\cos\psi I_{1r}(z) - \sin\psi I_{1i}(z))\right] + \right.$$

$$+ A_2\left[\cos 2\psi I_{0i}(z) + \sin 2\psi I_{0r}(z) - \frac{1-\nu}{r}(\cos\psi I_{1i}(z) + \sin\psi I_{1r}(z))\right] -$$

$$\left. - \frac{q}{2\pi k \sin 2\psi}\left[\cos 2\psi K_{0i}(z) + \sin 2\psi K_{0r}(z) + \frac{1-\nu}{\mathrm{r}}(\cos\psi K_{1i}(z) + \sin\psi K_{1r}(z))\right]\right\}, \tag{60}$$

$$M^{\varphi\varphi}(r) = -\frac{D}{L^2}\left\{A_1\left[\nu\cos 2\psi I_{0r}(z) - \nu\sin 2\psi I_{0i}(z) + \frac{1-\nu}{r}(\cos\psi I_{1r}(z) - \sin\psi I_{1i}(z))\right] + \right.$$

$$+ A_2\left[\nu\cos 2\psi I_{0i}(z) + \nu\sin 2\psi I_{0r}(z) + \frac{1-\nu}{r}(\cos\psi I_{1i}(z) + \sin\psi I_{1r}(z))\right] -$$

$$\left. - \frac{q}{2\pi k \sin 2\psi}\left[\nu\cos 2\psi K_{0i}(z) + \nu\sin 2\psi K_{0r}(z) - \frac{1-\nu}{r}(\cos\psi K_{1i}(z) + \sin\psi K_{1r}(z))\right]\right\}, \tag{61}$$

$$N^r(r) = -\frac{D}{L^3}\left\{A_1[\cos 3\psi I_{1r}(z) - \sin 3\psi I_{1i}(z)] + \right.$$

$$\left. + A_2[\cos 3\psi I_{1i}(z) + \sin 3\psi I_{1r}(z)] + \frac{q}{2\pi k \sin 2\psi}[\cos 3\psi K_{1i}(z) + \sin 3\psi K_{1r}(z)]\right\} \tag{62}$$

for $0 \leqq \varkappa < 1$, and an abbreviated notation $z = \mathrm{r}\, e^{i\psi}$ is used.

The constants A_1 and A_2 are to be evaluated from systems of algebraic equations generated by the boundary conditions (i) $w(R) = 0$ and $w'(R) = 0$ for a clamped edge and (ii) $w(R) = 0$ and $M^{rr}(R) = 0$ for a simply supported edge, and by the general solutions (51)—(62). The prime denotes a derivative.

In the case of a free edge the three boundary conditions

$$w(R) = w_F(R)\,, \qquad M^{rr}(R) = 0\,, \qquad N^r(R) + Gw'(R) = Gw'_F(R)$$

should be satisfied by three constants A_1, A_2 and B entering the solutions (51)—(62) and

$$w_F(r) = BK_0(\gamma r)\,, \quad \text{for} \quad r > R\,. \tag{63}$$

In Table 1 the dependence of dimensionless quantities

$$\overline{M}^{rr}(r) = \frac{kL^2}{qD}M^{rr}(r)\,, \qquad \overline{M}^{\varphi\varphi}(r) = \frac{kL^2}{qD}M^{\varphi\varphi}(r)\,, \qquad \overline{N}^r(r) = \frac{kL^3}{qD}N^r(r)$$

on r is shown for a clamped circular plate resting on the two-parameter foundation. The results are given for the concentrated loading $q(\mathbf{r}) = q\delta(\mathbf{r})$ and four characteristic values of $\varkappa$ and $R = 1$, $\nu = 0.3$. Maximum values of all these quantities correspond to the Winkler foundation ($\varkappa = 0$).

Table 1

r	$\varkappa = 0$			$\varkappa = 0.5$			$\varkappa = 1$			$\varkappa = 1.5$		
	$\overline{M}^{rr}$	$\overline{M}^{\varphi\varphi}$	$\overline{N}^r$	$\overline{M}^{rr}$	$\overline{M}^{\varphi\varphi}$	$\overline{N}^r$	$\overline{M}^{rr}$	$\overline{M}^{\varphi\varphi}$	$\overline{N}^r$	$\overline{M}^{rr}$	$\overline{M}^{\varphi\varphi}$	$\overline{N}^r$
0.1	0.157	0.213	−1.591	0.149	0.205	−1.573	0.142	0.197	−1.557	0.136	0.191	−1.541
0.2	0.086	0.142	−0.794	0.079	0.134	−0.769	0.074	0.128	−0.748	0.069	0.122	−0.727
0.3	0.044	0.099	−0.528	0.039	0.094	−0.501	0.036	0.088	−0.477	0.033	0.084	−0.456
0.4	0.015	0.070	−0.395	0.012	0.065	−0.368	0.011	0.061	−0.344	0.009	0.058	−0.323
0.5	−0.008	0.047	−0.315	−0.008	0.044	−0.289	−0.008	0.041	−0.268	−0.009	0.039	−0.248
0.6	−0.026	0.029	−0.262	−0.025	0.026	−0.239	−0.024	0.025	−0.220	−0.023	0.023	−0.203
0.7	−0.042	0.013	−0.224	−0.039	0.012	−0.206	−0.037	0.011	−0.190	−0.034	0.011	−0.176
0.8	−0.056	−0.0008	−0.196	−0.052	−0.0006	−0.183	−0.048	−0.0005	−0.172	−0.045	−0.0004	−0.162
0.9	−0.068	−0.013	−0.174	−0.063	−0.012	−0.167	−0.059	−0.011	−0.161	−0.056	−0.010	−0.156
1.0	−0.078	−0.024	−0.157	−0.074	−0.022	−0.157	−0.070	−0.021	−0.157	−0.067	−0.019	−0.157

b) BIE solution

In this part the general BIE solution of the Sec. 3c) will be applied to the circular plate subjected to the concentrated load $q(\boldsymbol{r}) = q\delta(\boldsymbol{r})$ with the clamped boundary.

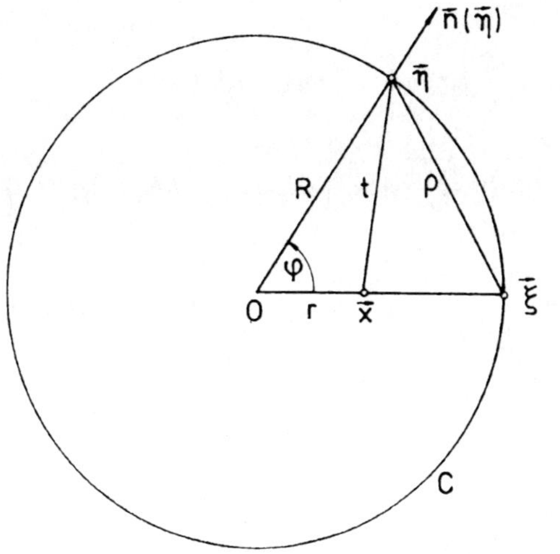

Fig. 1. Sketch of circular plate

According to rotational symmetry of the problem the unknown values $\nabla^2 w$ and $\partial\nabla^2 w/\partial \boldsymbol{n}$ are constant. The BIE (44) and (45) are converted into a system of algebraic equations

$$\frac{\partial\nabla^2\overline{w}}{\partial \boldsymbol{n}} I_1 + \nabla^2\overline{w} I_2 = B_1\,, \qquad \frac{\partial\nabla^2\overline{w}}{\partial \boldsymbol{n}} I_3 + \nabla^2\overline{w} I_4 = B_2 \tag{64}$$

where a dimensionless quantity $\overline{w}$ is defined by $\overline{w} = kq/w$. The integrals $I_1, \dots, I_4$, and the right-hand sides B_1 and B_2 are given for several $\varkappa$ as follows

$$\left.\begin{aligned}
I_1 &= \int_0^\pi [K_0(a\varrho) - K_0(b\varrho)]\,\mathrm{d}\varphi\,, \qquad I_2 = \int_0^\pi \sin\frac{\varphi}{2}\,[aK_1(a\varrho) - bK_1(b\varrho)]\,\mathrm{d}\varphi\,,\\
I_3 &= \int_0^\pi [a^2K_0(a\varrho) - b^2K_0(b\varrho)]\,\mathrm{d}\varphi\,, \qquad I_4 = \int_0^\pi \sin\frac{\varphi}{2}\,[a^3K_1(a\varrho) - b^3\,K_1(b\varrho)]\mathrm{d}\varphi - \frac{\pi}{R}\sqrt{\varkappa^2 - 1}\,,\\
B_1 &= \frac{1}{2R}[K_0(aR) - K_0(bR)]\,, \qquad B_2 = \frac{1}{2R}[a^2K_0(aR) - b^2K_0(bR)]
\end{aligned}\right\} \tag{65}$$

for $\varkappa > 1$,

$$\left.\begin{aligned}
I_1 &= \int_0^\pi \varrho K_1(\varrho)\,\mathrm{d}\varphi\,, \qquad I_2 = \int_0^\pi \sin\frac{\varphi}{2}\,\varrho K_0(\varrho)\,\mathrm{d}\varphi\,,\\
I_3 &= \int_0^\pi K_0(\varrho)\,\mathrm{d}\varphi\,, \qquad I_4 = \int_0^\pi \sin\frac{\varphi}{2}\,K_1(\varrho)\,\mathrm{d}\varphi - \frac{\pi}{2R},\\
B_1 &= \frac{1}{2}\,K_1(R)\,, \qquad B_2 = \frac{1}{2R}\,K_0(R)
\end{aligned}\right\} \tag{66}$$

for $\varkappa = 1$, and

$$\left.\begin{aligned}
I_1 &= \int_0^\pi \operatorname{Im} K_0(\varrho\,\mathrm{e}^{i\psi})\,\mathrm{d}\varphi\,, \qquad I_2 = \int_0^\pi \sin\frac{\varphi}{2}\operatorname{Im}\,[\mathrm{e}^{i\psi}\,K_1(\varrho\,\mathrm{e}^{i\psi})]\,\mathrm{d}\varphi\\
I_3 &= \int_0^\pi \operatorname{Im}\,[\mathrm{e}^{i2\psi}\,K_0(\varrho\,\mathrm{e}^{i\psi})]\,\mathrm{d}\varphi\,, \qquad I_4 = \int_0^\pi \sin\frac{\varphi}{2}\operatorname{Im}\,[\mathrm{e}^{i3\psi}\,K_1(\varrho\,\mathrm{e}^{i\psi})]\,\mathrm{d}\varphi - \frac{\pi}{2R}\sin 2\psi\,,\\
B_1 &= \frac{1}{2R}\operatorname{Im} K_0(R\,\mathrm{e}^{i\psi})\,, \qquad B_2 = \frac{1}{2R}\operatorname{Im}\,[\mathrm{e}^{i2\psi}\,K_0(R\,\mathrm{e}^{i\psi})]
\end{aligned}\right\} \tag{67}$$

for $0 \leqq \varkappa < 1$, where $\varrho = 2R\sin\dfrac{\varphi}{2}$.

Finally $w(r)$ are calculated by the following equations

$$\overline{w}(r) = \frac{K_0(br) - K_0(ar)}{4\pi\sqrt{\varkappa^2 - 1}} + \frac{R}{2\pi\sqrt{\varkappa^2 - 1}} \frac{\partial\nabla^2\overline{w}}{\partial \boldsymbol{n}} \int_0^\pi [K_0(at) - K_0(bt)]\, d\varphi +$$

$$+ \frac{R}{2\pi\sqrt{\varkappa^2 - 1}} \nabla^2\overline{w} \int_0^\pi \left[\frac{R\cos\varphi - r}{t}\cos\varphi + \frac{R}{t}\sin^2\varphi\right] [aK_1(at) - bK_1(bt)]\, d\varphi\,, \tag{68}$$

$$\overline{w}(0) = \frac{1}{2\pi\sqrt{\varkappa^2 - 1}} \left\{\frac{1}{2}\ln\frac{a}{b} + \pi R[K_0(aR) - K_0(bR)] \frac{\partial\nabla^2\overline{w}}{\partial \boldsymbol{n}} + \pi R[aK_1(aR) - bK_1(bR)]\, \nabla^2\overline{w}\right\} \tag{69}$$

for $\varkappa > 1$,

$$\overline{w}(r) = \frac{r}{4\pi} K_1(r) - \frac{R}{2\pi} \frac{\partial\nabla^2\overline{w}}{\partial \boldsymbol{n}} \int_0^\pi tK_1(t)\, d\varphi - \frac{R}{2\pi} \nabla^2\overline{w} \int_0^\pi [(R\cos\varphi - r)\cos\varphi + R\sin^2\varphi]\, K_0(t)\, d\varphi\,, \tag{70}$$

$$\overline{w}(0) = \frac{1}{4\pi} - \frac{R^2}{2} K_1(R) \frac{\partial\nabla^2\overline{w}}{\partial \boldsymbol{n}} - \frac{R^2}{2} K_0(R)\, \nabla^2\overline{w} \tag{71}$$

for $\varkappa = 1$, and

$$\overline{w}(r) = \frac{1}{2\pi\sin 2\psi} \left\{-\mathrm{Im}\, K_0(r\, e^{i\psi}) + 2R \frac{\partial\nabla^2\overline{w}}{\partial \boldsymbol{n}} \int_0^\pi \mathrm{Im}\, K_0(t\, e^{i\psi})\, d\varphi +\right.$$

$$\left. + 2R\nabla^2\overline{w} \int_0^\pi \left[\frac{R\cos\varphi - r}{t}\cos\varphi + \frac{R}{t}\sin^2\varphi\right] \mathrm{Im}\, [e^{i\psi} K_1(t\, e^{i\psi})]\, d\varphi\right\}, \tag{72}$$

$$\overline{w}(0) = \frac{1}{2\pi\sin 2\psi} \left\{\psi + 2\pi R \frac{\partial\nabla^2\overline{w}}{\partial \boldsymbol{n}} \mathrm{Im}\, K_0(R\, e^{i\psi}) + 2\pi R\nabla^2\overline{w}\, \mathrm{Im}\, [e^{i\psi} K_1(R\, e^{i\psi})]\right\} \tag{73}$$

für $0 \leq \varkappa < 1$, where $t = (R^2 + r^2 - 2rR\cos\varphi)^{1/2}$.

The dimensionless deflection $\overline{w}(r)$ of a clamped circular plate subjected to the concentrated load $q(\boldsymbol{r}) = q\delta(\boldsymbol{r})$ has been calculated either analytically and numerically by the BIEM. The results are compared in Table 2. The dependence of $\overline{w}$ on r for various $\varkappa$ is quantitatively the same. The maximum deflections correspond to the Winkler foundation ($\varkappa = 0$).

Table 2

r	$\overline{w}$ for $\varkappa = 0$		$\overline{w}$ for $\varkappa = 0.5$		$\overline{w}$ for $\varkappa = 1$		$\overline{w}$ for $\varkappa = 1.5$	
	analytical	BIE	analytical	BIE	analytical	BIE	analytical	BIE
0	0.0197	0.0188	0.0185	0.0179	0.0176	0.0180	0.0167	0.0167
0.1	0.0186	0.0176	0.0175	0.0168	0.0165	0.0172	0.0157	0.0152
0.2	0.0164	0.0154	0.0154	0.0147	0.0145	0.0151	0.0137	0.0132
0.3	0.0137	0.0127	0.0128	0.0121	0.0120	0.0125	0.0113	0.0108
0.4	0.0108	0.0098	0.0101	0.0094	0.0094	0.0099	0.0089	0.0084
0.5	0.0079	0.0071	0.0074	0.0068	0.0069	0.0073	0.0065	0.0061
0.6	0.0054	0.0046	0.0050	0.0044	0.0047	0.0049	0.0044	0.0040
0.7	0.0032	0.0025	0.0029	0.0025	0.0028	0.0029	0.0026	0.0023
0.8	0.0015	0.0010	0.0014	0.0010	0.0013	0.0014	0.0012	0.0009
0.9	0.0004	0.0001	0.0003	0.0001	0.0003	0.0004	0.0003	0.0002
1.0	0	0	0	0	0	0	0	0

References

1 VLASOV, V. Z.; LEONTIEV, N. N., Balki, plity i oboločki na uprugom osnovanii, Fizmatgiz, Moskva 1960.
2 SELVADURAI, A. P. S., Elastic analysis of soil-foundation interaction, Elsevier, Amsterdam-Oxford-New York 1979.
3 KERR, A. D., Elastic and viscoelastic foundation models, J. Appl. Mech. 31, 491–498 (1964).
4 HANSEN, E. B., Numerical solution of integro-differential and singular equations for plate bending problems, J. Elasticity 6 (1976), 39–56.
5 BATEMAN, H.; ERDÉLYI, A., Higher transcendental functions vol. 2, Mc Graw-Hill Book Company, Inc., New York 1953.
6 FLÜGGE, W., Tensor analysis and Continuum mechanics, Springer Verlag, 1973.

Received August 5, 1982

Anschrift: Correspondent member J. BALAŠ DrSc, Dr. V. SLÁDEK, and Ing. J. SLÁDEK CSc, Institute of Construction and Architecture, Slovak Academy of Sciences, 84220 Bratislava, Czechoslovakia

ZAMM 64, 147 – 153 (1984)

J. A. ANDREJEWA / R. KLÖTZLER

Zur analytischen Lösung geometrischer Optimierungsaufgaben mittels Dualität bei Steuerungsproblemen. Teil II

Im Teil I der gleichlautenden Arbeit wurden Grundlagen zur Dualität bei Steuerungsproblemen bereitgestellt und auf zwei geometrische Optimierungsprobleme angewandt: Die Bestimmung ebener konvexer Bereiche größten bzw. kleinsten Umfangs bei vorgegebener Dicke Δ und vorgeschriebenem Durchmesser D. Im vorliegenden Teil II schließt sich unter gleichen Nebenbedingungen die Behandlung zweier weiterer Beispiele an: Das Problem des ebenen konvexen Bereichs größten Flächeninhalts und des konvexen Körpers größter Oberfläche.

In Part I of this article some bases were stated for the duality in optimal control. They were applicated to two geometrical problems of optimization: the investigation of plane convex domains with greatest or smallest perimeter, resp., under given thickness Δ and diameter D. In the present Part II the consideration of two further examples is continued by the same side-conditions: the problem of the plane convex domain with greatest area and the problem of the convex body with greatest surface area.

В первой части этой работы мы положили основы для двойственности в теории оптимального управления и применили их к двум геометрическим проблемам оптимизации: К нахождению плоской выпуклой области максимального или минимального объёма при данной ширины Δ и данному диаметру D. В настоящей части II рассматриваются при тем же условиям два допольнительных примера: Проблема нахождения плоской выпуклой области максимальной площади и выпуклого тела максимальной поверхности.

Einleitung

Dieser Teil II bildet die unmittelbare Fortsetzung des gleichlautenden Artikels Teil I [1], in dem die methodischen Grundlagen zur Lösung spezieller Optimierungsaufgaben für konvexe Figuren bereitgestellt wurden. Wir stützen uns im vorliegenden Artikel voll auf diese Ergebnisse, so daß dem aufmerksamen Leser die gleichzeitige Heranziehung des Teiles I sehr zu empfehlen ist. Dennoch seien zur Erleichterung des Verständnisses einige Erklärungen zu Bezeichnungen wiederholt, die für die Formulierung der nachstehenden Anwendungsbeispiele wesentlich sind.

$\mathfrak{A}_n$ bezeichnet die Klasse aller beschränkten konvexen abgeschlossenen Punktmengen $\mathfrak{B}$ des E^n. $\mathfrak{A}_n(D, \Delta)$ ist die Teilmenge aller $\mathfrak{B}$ von $\mathfrak{A}_n$ mit vorgegebenem Durchmesser D und vorgegebener Dicke Δ. h stellt die Stützfunktion eines $\mathfrak{B}$ dar.

Die Numerierung der Kapitel, Formeln und Abbildungen erfolgt in Fortsetzung der Zählung von Teil I. Das Literaturverzeichnis wird hingegen für Teil II separat geführt.

5. Fortführung der Beispiele

c) Das Problem des größten Flächeninhalts F in der Klasse $\mathfrak{A}_2(D, \Delta)$

Nach (11a), (12), (15) und (19) lautet eine analytische Formulierung dieses geometrischen Optimierungsproblems

$$F = \frac{1}{2} \int_0^{2\pi} (-h'^2 + h^2) \, d\varphi \to \sup \quad \text{im} \quad W_1^2(0, 2\pi) \tag{31}$$

unter den Nebenbedingungen $\Delta \leqq h(\varphi) + h(\varphi + \pi) \leqq D$, $h'' + h \geqq 0$ und den Randbedingungen $h(0) = h(\pi) = \Delta/2$.

Eigentlich müßte in die Zustandsbeschränkungen von (31) noch die Bedingung $\operatorname{Max}_{[0,\pi]} [h(\varphi) + h(\varphi + \pi)] = D$ aufgenommen werden. Wir werden jedoch sehen, daß für die optimale Lösung des zugeordneten Steuerungsproblems zu (31) jene Eigenschaft von selbst erfüllt ist. Unter dem gleichen Aspekt erlauben wir uns zur Vereinfachung, beim Übergang zum zugeordneten Steuerungsproblem die Bedingung $h'' + h \geqq 0$ zu vernachlässigen. Mit der Bezeichnung $x^1(\varphi) := h(\varphi)$ und $x^2(\varphi) := h(\varphi + \pi)$ sowie $u^i(\varphi) := \dot{x}^i(\varphi)$, $i = 1, 2$, für alle $\varphi \in [0, \pi]$ entsteht so die zugeordnete (abgeschwächte) Steuerungsaufgabe

$$J(x, u) = \frac{1}{2} \int_0^{\pi} \sum_{i=1}^{2} [(u^i)^2 - (x^i)^2] \, dt \to \inf \tag{31'}$$

unter den Zustandsgleichungen $\dot{x}^i = u^i$, $i = 1, 2$, den Zustandsbeschränkungen $\Delta \leqq x^1(t) + x^2(t) \leqq D$ und den Randbedingungen $x^1(0) = x^1(\pi) = x^2(0) = x^2(\pi) = \Delta/2$. Steuerbeschränkungen entfallen hier.

Die Pontrjaginsche Funktion des Problems (31′) ist

$$H(t, \xi, v, y) = -\frac{1}{2} \sum_{i=1}^{2} [(v^i)^2 - (\xi^i)^2] + \sum_{i=1}^{2} y_i v^i$$

und die Hamiltonfunktion errechnet sich daraus zu

$$\mathcal{H}(t, \xi, y) = \sup_v H(t, \xi, v, y) = \frac{1}{2} \sum_{i=1}^{2} [(\xi^i)^2 + (y_i)^2]$$

mit $v_{\max} = y$. Wir bemühen uns, Satz 2 auf Problem (31′) anzuwenden unter Benutzung des quadratischen Ansatzes

$$S(t, \xi) = a(t) + \sum_{i=1}^{2} y_i(t)\left(\xi^i - x^i(t)\right) + \tfrac{1}{2}\,\psi(t)\left(\xi^i - x^i(t)\right)^2, \tag{32}$$

wobei hier x die noch unbekannte optimale Zustandsfunktion zu (31′) bezeichnet.

Damit lautet Forderung (2): Abgesehen von endlich vielen Ausnahmestellen ist auf $[0, \pi] \times E^2$

$$\begin{aligned} \Lambda(t, \xi) &:= S_t(t, \xi) + \mathcal{H}\left(t, \xi, S_\xi(t, \xi)\right) = \\ &= \dot a(t) + \sum_{i=1}^{2} \{\dot y_i(t)\left(\xi^i - x^i(t)\right) - y_i(t)\,\dot x^i(t) + \tfrac{1}{2}\,\dot\psi(t)\left(\xi^i - x^i(t)\right)^2 - \\ &\quad - \psi(t)\left(\xi^i - x^i(t)\right)\dot x^i(t) + \tfrac{1}{2}\left[(\xi^i)^2 + S^2_{\xi^i}\right]\} \leqq 0 \end{aligned} \tag{33a}$$

unter der Nebenbedingung $\Delta \leqq \xi^1 + \xi^2 \leqq D$.

Wir suchen gemäß Satz 2 das S so zu bestimmen, daß in den Punkten $\left((t, x(t)\right)$ die Bedingung (5) erfüllt ist. Danach muß gemäß (5b) für $\xi = x(t)$ speziell

$$\Lambda(t, x(t)) = 0 \quad \text{in } [0, \pi] \text{ (bis auf endlich viele } t) \tag{33b}$$

gelten. Somit nimmt $\Lambda(t, \xi)$ bei festem $t \in [0, \pi]$ seinen Maximalwert im Zustandsbereich für $\xi = x(t)$ an. Notwendig dafür ist die Gültigkeit der Kuhn-Tucker-Bedingung

$$\begin{aligned} &\Lambda_{\xi^i}(t, x(t)) + \nu(t) - \mu(t) = 0\,, \qquad i = 1, 2 \quad \text{mit} \quad \nu \geqq 0\,, \qquad \mu \geqq 0\,, \\ &\nu(t)\,(\Delta - x^1(t) - x^2(t)) = 0\,, \qquad \mu(t)\,(x^1(t) + x^2(t) - D) = 0 \end{aligned}$$

in allen Punkten t, wo y und ψ differenzierbar sind. Ausgerechnet bedeutet das

$$\left.\begin{aligned} &\dot y_i + x^i + \nu - \mu = 0\,, \qquad i = 1, 2\,, \quad \text{mit} \quad \nu \geqq 0\,, \qquad \mu \geqq 0\,, \\ &\nu(t)\,(\Delta - x^1(t) - x^2(t)) = 0\,, \qquad \mu(t)\,(x^1(t) + x^2(t) - D) = 0\,. \end{aligned}\right\} \tag{34a}$$

Indem wir (x, u) als optimalen Prozeß zu (31′) auffassen, muß zugleich wegen (5a)

$$\dot x^i = u^i = y_i\,, \qquad i = 1, 2\,, \tag{34b}$$

gelten. $a(t)$ berechnen wir aus (33b). Gemäß (3) ist mit $\Omega_j = (\tau_{j-1}, \tau_j), \tau_0 = 0, \tau_\varkappa = \pi$, analog zu (26)

$$\hat L(S) = \inf_{\zeta \in \hat Q} \{S(\pi, \zeta(\pi)) - S(0, \zeta(0)) + \sum_{j=1}^{\varkappa-1} [S(\tau_j - 0, \zeta(\tau_j)) - S(\tau_j + 0, \zeta(\tau_j))]\}\,. \tag{35}$$

Bei Annahme eines Minimums in diesem Ausdruck für $\zeta(\tau_j) = x(\tau_j)$ gemäß (5c) erhalten wir nach abermaliger Anwendung des Kuhn-Tucker-Theorems als notwendige Bedingung

$$\left.\begin{aligned} &y_i(\tau_j - 0) - y_i(\tau_j + 0) - \nu_j + \mu_j = 0\,, \qquad i = 1, 2\,, \quad \text{mit} \quad \nu_j \geqq 0\,, \\ &\mu_j \geqq 0\,, \qquad \nu_j(\Delta - x^1(\tau_j) - x^2(\tau_j)) = 0\,, \qquad \mu_j(x^1(\tau_j) + x^2(\tau_j) - D) = 0\,. \end{aligned}\right\} \tag{36}$$

Somit können nur dort Unstetigkeiten von $y = (y_1, y_2)$ auftreten, wo $x^1(\tau_j) + x^2(\tau_j) = \Delta$ oder D ist.

Um zu erreichen, daß die Bedingungen (34a) und (36) zugleich hinreichend für das Erfülltsein von (2) und (5b) (und damit von (33)) bzw. (5c) sind, setzen wir $\psi(t) = \cot t$ in $(0, \pi)$. Damit ist einerseits ψ stetig in diesem Intervall, so daß im rechten Ausdruck von (35) keine quadratischen Glieder bzgl. ζ mehr auftreten. Zum anderen verschwinden damit zugleich in (33) die quadratischen Terme in ξ wegen $1 + \dot\psi + \psi^2 = 0$.

Die (einparametrige) Gesamtheit aller Lösungen der Bedingungen (34) und (36) läßt sich leicht errechnen und durch das folgende Funktionensystem beschreiben:

Mit $\alpha = \arccos(\Delta/D)$ ist zu beliebigen Parametern $c_1 = -c_2$

$$\text{für } 0 \leqq t \leqq \alpha: x^i(t) = \frac{D}{2}\cos(t - \alpha) + c_i \sin t\,, \qquad u^i(t) = -\frac{D}{2}\sin(t - \alpha) + c_i \cos t = y_i(t)\,, \qquad i = 1, 2;$$

$$\text{für } \alpha \leqq t \leqq \frac{\pi}{2}: x^i(t) = \frac{D}{2} + c_i \cos t\,, \qquad u^i(t) = c_i \cos t = y_i(t)\,, \qquad i = 1, 2;$$

$$x^i\left(\frac{\pi}{2} + \tau\right) = x^i\left(\frac{\pi}{2} - \tau\right), \qquad u^i\left(\frac{\pi}{2} + \tau\right) = -u^i\left(\frac{\pi}{2} - \tau\right), \qquad y_i\left(\frac{\pi}{2} + \tau\right) = -y_i\left(\frac{\pi}{2} - \tau\right),$$

$$i = 1, 2 \quad \text{für} \quad 0 \leqq \tau \leqq \frac{\pi}{2}.$$

$a(t)$ berechnen wir — wie schon oben bemerkt — aus (33b) durch Integration über t.

In der Abb. 5 wird diese Lösung für den Spezialfall $c_1 = c_2 = 0$ graphisch dargestellt. Für jedes zulässige fixierte Parameterpaar c_1, c_2 wird durch diese Lösung ein Prozeß (x, u) bestimmt, der bzgl. (31′) optimal ist infolge von Satz 2. Da dieser Prozeß für $t \neq \alpha$ aber auch die Bedingung $\ddot x^i + x^i \geqq 0$, $i = 1, 2$, erfüllt, liefert er über die Erklärung $h(\varphi) := x^1(\varphi)$, $h(\varphi + \pi) := x^2(\varphi)$ zu $\varphi \in [0, \pi]$ zugleich eine (nicht-reguläre) Stützfunktion h, für die $-J(x, u) = F(h) = \sup F$ bzgl. des Problems (31) gilt. Jedem dieser so erzeugten h entspricht geometrisch über die Beziehungen (10) und (11a) ein konvexer Bereich $\mathfrak{B}$ als optimale Lösung unseres Problems c). Alle diese optimalen Bereiche gehen auseinander durch eine Parallelverschiebung in Richtung $\varphi = \pi/2$ hervor, sie sind also kongruent.

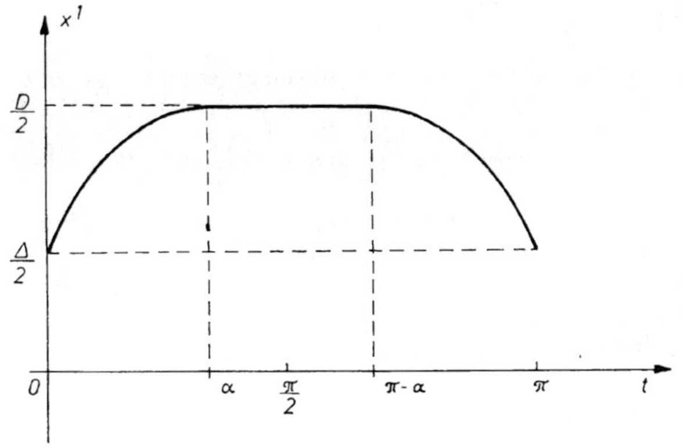

Abb. 5

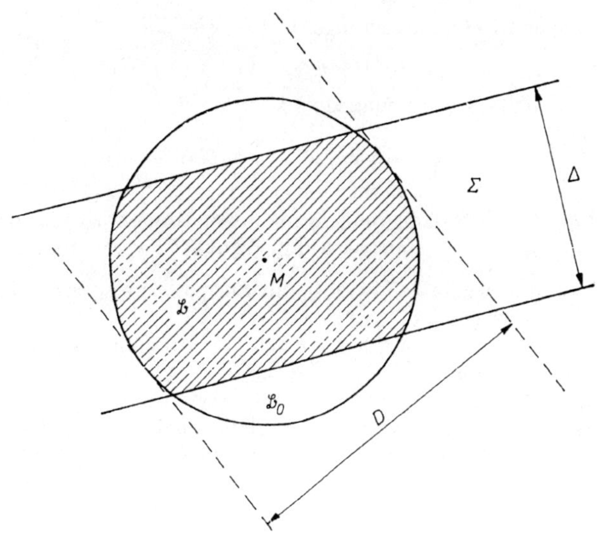

Abb. 6

Elementar-geometrisch ausgedrückt lautet unser Resultat:

Behauptung 3: *Eine optimale Lösung des Problems c) ist ein Bereich $\mathfrak{B}$, der als Durchschnitt eines Parallelstreifens Σ der Dicke Δ mit einem Kreis $\mathfrak{B}_0$ des Durchmessers D entsteht, wenn der Mittelpunkt M von $\mathfrak{B}_0$ von den Begrenzungsgeraden zu Σ den gleichen Abstand besitzt.* (Man vergleiche dazu auch entsprechende Aussagen von [3] Aufg. 76b sowie Abb. 6).

Bemerkung 1: Das in Behauptung 3 ausgesprochene Resultat widerspiegelt eine bekannte geometrische Aussage, die schon von T. KUBOTA [4] gefunden wurde (über Symmetrisierungsmethoden). Sie gibt Anlaß zu der isoperimetrischen Ungleichung

$$F \leqq \frac{1}{2}\left[\Delta \sqrt{D^2 - \Delta^2} + D^2 \arcsin \frac{\Delta}{D}\right]^{1)}.$$

Die bisher noch offene Frage der Eindeutigkeit der optimalen Lösung von Problemen c) (bis auf Kongruenz) haben wir mit der obigen analytischen Herleitung der Lösungen zugleich positiv beantworten können.

Bemerkung 2: Mit $\psi(t) = \cot t$ ist S gemäß (32) in dem hier zugrundegelegten Zustandsbereich

$$\overline{G} = \{(t, \xi^1, \xi^2) \in E^3 \mid 0 \leqq t \leqq \pi,\ \Delta \leqq \xi^1 + \xi^2 \leqq D\}$$

nicht stückweise stetig differenzierbar, wie es Satz 2 fordert. Diese logische Lücke kann geschlossen werden, indem wir beachten, daß wegen der geometrischen Bedeutung von h' in Aufgabe (31) ohne Einschränkung der Allgemeinheit $|h'| \leqq D$ gefordert werden kann (vgl. [2] S. 66), sofern wir auch noch die Nebenbedingung $h \geqq 0$ aufnehmen. Diese kann aber zu einem zu beschreibenden konvexen Bereich $\mathfrak{B}$ stets erfüllt werden, wenn man den Pol des verwendeten Polarkoordinatensystems aus $\mathfrak{B}$ selbst wählt. Als Zustandsraum interessiert dann zum zugeordneten Steuerungsproblem (31') lediglich der Teilbereich $G_0 \subset G$ als Schnitt von G mit den Kegeln

$$\left|\xi^i - \frac{\Delta}{2}\right| \leqq Dt \quad \text{und} \quad \left|\xi^i - \frac{\Delta}{2}\right| \leqq D(\pi - t)\,, \qquad i = 1, 2\,.$$

In diesem Bereich G_0 ist aber wegen $x^i(0) = \frac{\Delta}{2} = x^i(\pi)$, $i = 1, 2$, sowohl $\psi(t) \sum_{i=1}^{2} (\xi^i - x^i(t))^2$ als auch

$$\dot\psi(t) \sum_{i=1}^{2} (\xi^i - x^i(t))^2 \quad \text{und} \quad \psi(t) \sum_{i=1}^{2} (\xi^i - \dot x^i(t))$$

beschränkt, so daß S in G_0 tatsächlich im Sinne von Satz 2 zulässig ist.

d) Das Problem der größten Oberfläche O in der Klasse $\mathfrak{A}_3(D, \Delta)$

Nach (11b), (13), (14), (15) und (20) lautet eine analytische Formulierung dieses geometrischen Optimierungsproblems

$$O = \int_A \left[h^2 - \frac{1}{2} h_\varphi^2 - \frac{1}{2\sin^2\varphi} h_\theta^2\right] |\sin\varphi| \, d\theta \, d\varphi \to \sup \tag{37}$$

mit

$$A = \{(\theta, \varphi) \in E^2 \mid 0 \leqq \theta \leqq 2\pi,\ 0 \leqq \varphi \leqq \pi\}$$

[1]) In [2] S. 81 Formel (3) ist durch einen Druckfehler der Faktor 1/2 entfallen.

unter den Nebenbedingungen (13) sowie

$$\Delta \leqq h(\theta, \varphi) + h(\pi + \theta, \pi - \varphi) \leqq D$$

und den Randbedingungen

$$h(0, \varphi) = h(2\pi, \varphi)\,, \qquad h(\theta, 0) = h(\pi + \theta, \pi) = \frac{\Delta}{2}, \qquad h\left(\pi, \frac{\pi}{2}\right) = h\left(0, \frac{\pi}{2}\right).$$

Ohne Einschränkung der Allgemeinheit können wir die zweite und dritte Randbedingung durch geeignete Lage jedes $\mathfrak{B}$ aus $\mathfrak{A}_3(D, \Delta)$ im sphärischen Koordinatensystem realisieren. Auf die Bedingung $\operatorname{Max}_{\Lambda} [h(\theta, \varphi) + h(\pi + \theta, \pi - \varphi)] = D$ verzichten wir, indem wir wie im Beispiel a) erwarten, daß sich für optimale Lösungen des Problems d) von selbst diese Eigenschaft einstellt.

Im weiteren werden wir zur Vereinfachung sogar auf die volle Nebenbedingung (13) verzichten und lediglich statt dessen die Forderung $R = h_{\varphi\varphi} + h \geqq 0$ verwenden, d. h., wir werden in Analogie zu vorangehenden Beispielen unsere Aufgabe (37) in einem erweiterten zulässigen Bereich studieren. Es wird sich dabei zeigen, daß optimale Lösungen dieses erweiterten Problems auch Lösungen von Beispiel d) entsprechen.

Wir formulieren dieses erweiterte Problem unter Verwendung der Hilfsfunktionen

$$\left.\begin{aligned} x^1(\theta, \varphi) &:= h\left(\theta + \frac{\pi}{2}, \varphi\right) + h\left(\theta + \frac{3}{2}\pi, \pi - \varphi\right) && \text{in } A_1 \text{ mit} \\ x^2(\theta, \varphi) &:= h\left(\theta + \frac{\pi}{2}, \varphi\right) - h\left(\theta + \frac{3}{2}\pi, \pi - \varphi\right) && A_1 := \left[-\frac{\pi}{2}, \frac{\pi}{2}\right] \times \left[0, \frac{\pi}{2}\right] \end{aligned}\right\} \tag{38a}$$

bzw.

$$\left.\begin{aligned} x^1(\theta, \varphi) &:= h\left(\theta + \frac{\pi}{2}, \varphi + \pi\right) + h\left(\theta + \frac{3}{2}\pi, -\varphi\right) && \text{in } A_2 \text{ mit} \\ x^2(\theta, \varphi) &:= h\left(\theta + \frac{\pi}{2}, \varphi + \pi\right) - h\left(\theta + \frac{3}{2}\pi, -\varphi\right) && A_2 := \left[-\frac{\pi}{2}, \frac{\pi}{2}\right] \times \left[-\frac{\pi}{2}, 0\right]. \end{aligned}\right\} \tag{38b}$$

Außerdem setzen wir

$$x^3 := x^1_\varphi\,, \qquad x^4 := x^2_\varphi\,, \qquad x^5 := x^1_\theta\,, \qquad x^6 := x^2_\theta\,, \qquad u^1 := x^3_\varphi + x^1\,. \tag{39}$$

Wir erhalten so als erweitertes Problem zu (37)

$$J(x, u) = \int\limits_{A_1 \cup A_2} \left\{\frac{(x^5)^2 + (x^6)^2}{2\,|\sin\varphi|} + \frac{(x^3)^2 + (x^4)^2}{2} |\sin\varphi| - [(x^1)^2 + (x^2)^2]\,|\sin\varphi|\right\} d\theta\, d\varphi \to \inf \tag{40a}$$

unter den Zustandsgleichungen

$$x^1_\varphi = x^3\,, \qquad x^1_\theta = x^5\,, \qquad x^2_\theta = x^6\,, \qquad x^2_\varphi = x^4\,, \qquad x^3_\varphi = -x^1 + u^1\,, \tag{40b}$$

den Zustandsbeschränkungen

$$\Delta \leqq x^1(\theta, \varphi) \leqq D\,, \tag{40c}$$

den Steuerbeschränkungen

$$u^1(\theta, \varphi) \geqq 0 \tag{40d}$$

und den Nebenbedingungen[2])

$$x^1(\theta, 0) = \Delta\,, \qquad x^2(\theta, 0) = 0\,, \qquad x^2\left(\pm\frac{\pi}{2}, \frac{\pi}{2}\right) = 0\,, \qquad x^5(\theta, 0) = x^6(\theta, 0) = 0 \tag{40e}$$

sowie

$$x^2\left(\frac{\pi}{2}, \varphi\right) = -x^2\left(-\frac{\pi}{2}, -\varphi\right). \tag{40f}$$

Wir werden uns bemühen, auf Problem (40) den Satz 2 anzuwenden. Das wird möglich sein, obwohl die Menge der Zustandsgleichungen zu den Zustandsfunktionen $x^1, \ldots, x^6$ nicht vollständig ist, indem den Ableitungen x^3_θ als auch x^j_θ und x^j_φ für $j = 4, 5, 6$ keine Bedingungen auferlegt werden. Man muß dann die vorangestellte Theorie lediglich so modifiziert anwenden, daß zu den nichtauftretenden $x^{j}_{t^\alpha}$ die entsprechenden $S^\alpha_{\xi^j} \equiv 0$ sind. In unserem Beispiel bedeutet das also, daß mit $\theta = t^1$, $\varphi = t^2$ die Ansätze für S^α so gewählt werden müssen, daß $S^1_{\xi^3} \equiv 0$ und $S^\alpha_{\xi^j} \equiv 0$ für $\alpha = 1, 2$ und $j = 4, 5, 6$ sind.

Die Pontrjaginsche Funktion des Problems (40) lautet (mit $y^1_3 = 0$, $y^\alpha_j = 0$ für $\alpha = 1, 2$; $j = 4, 5, 6$)

$$\begin{aligned} H(t, \xi, v, y) = {} & [(\xi^1)^2 + (\xi^2)^2]\,|\sin t^2| - \frac{(\xi^3)^2 + (\xi^4)^2}{2} |\sin t^2| - \\ & - \frac{(\xi^5)^2 + (\xi^6)^2}{2|\sin t^2|} + y^1_1\xi^5 + y^2_1\xi^3 + y^1_2\xi^6 + y^2_2\xi^4 - y^2_3\xi^1\,. \end{aligned} \tag{41}$$

[2]) Im Sinne der Nebenbedingungen $b(x) = 0$ auf $\mathfrak{R}$ gemäß (1).

Die zugeordnete Hamiltonfunktion berechnet sich daraus wegen (40d) zu

$$\mathcal{H}(t, \xi, y) = [(\xi^1)^2 + (\xi^2)^2]\,|\sin t^2| - \frac{(\xi^3)^2 + (\xi^4)^2}{2}\,|\sin t^2| -$$

$$- \frac{(\xi^5)^2 + (\xi^6)^2}{2|\sin t^2|} + y_1^1\xi^5 + y_1^2\xi^3 + y_2^1\xi^6 + y_2^2\xi^4 - y_3^2\xi^1, \tag{42}$$

falls $y_3^2 \leqq 0$ ist, wobei $v^1_{\max}(t, \xi, y)$ der Bedingung $v^1_{\max} y_3^2 = 0$ neben $v^1_{\max} \geqq 0$ genügt. Für $y_3^2 > 0$ ist $\mathcal{H}(t, \xi, y) = \infty$.

Für $S^\alpha(\alpha = 1, 2)$ machen wir folgende Ansätze

$$S^1(t, \xi) = \frac{\Phi(t)}{2}\,(\xi^2)^2\,, \tag{43a}$$

$$S^2(t, \xi) = a^2(t) + y_1^2(t)\,(\xi^1 - x^1(t)) + y_3^2(t)\,(\xi^3 - x^3(t)) + \frac{\psi_1(t)}{2}\,(\xi^1 - x^1(t))^2 + \frac{\psi_2(t)}{2}\,(\xi^2)^2 \tag{43b}$$

mit $y_3^2 \leqq 0$. Dabei bezeichnet x die noch zu errechnende optimale Zustandsfunktion zu (40) in Verbindung mit der Steuerung u. Mit diesen S^α lautet zu Problem (40) die Forderung (2):

$$\Lambda(t, \xi) = \frac{\Phi_{t^1}}{2}\,(\xi^2)^2 + a^2_{t^2} + y^2_{1t^2}(\xi^1 - x^1) - y_1^2 x^1_{t^2} + y^2_{3t^2}(\xi^3 - x^3) - y_3^2 x^3_{t^2} + \frac{\psi_{1t^2}}{2}\,(\xi^1 - x^1)^2 -$$

$$- \psi_1(\xi^1 - x^1)\,x^1_{t^2} + \frac{\psi_{2t^2}}{2}\,(\xi^2)^2 + [(\xi^1)^2 + (\xi^2)^2]\,|\sin t^2| - \frac{(\xi^3)^2 + (\xi^4)^2}{2}|\sin t^2| - \frac{(\xi^5)^2 + (\xi^6)^2}{2\,|\sin t^2|} -$$

$$- y_3^2\xi^1 + [y_1^2 + \psi_1(\xi^1 - x^1)]\cdot \xi^3 + \Phi\xi^2\xi^6 + \psi_2\xi^2\xi^4 \leqq 0 \tag{44}$$

für $t \in A_1 \cup A_2$ und $\Delta \leqq \xi^1 \leqq D$.

Indem wir weiterhin S^α so wählen, daß es sämtlichen Bedingungen von (5) genügt, muß gemäß (5b) für $\xi = x(t)$ speziell $\Lambda(t, x(t)) = 0$ gelten für fast alle $t \in A_1 \cup A_2$. Somit nimmt $\Lambda(t, \xi)$ für fast alle $t \in A_1 \cup A_2$ seinen Maximalwert Null im Zustandsbereich für $\xi = x(t)$ an, so daß notwendig an dieser Stelle die Kuhn-Tucker-Bedingungen gelten müssen: Es existieren $\nu(t) \geqq 0$ und $\mu(t) \geqq 0$ mit

$$\Lambda_{\xi^1}(t, x(t)) + \nu(t) - \mu(t) = 0\,, \qquad \nu(t)\,(\Delta - x^1(t)) = 0\,, \qquad \mu(t)\,(D - x^1(t)) = 0$$

und

$$\Lambda_{\xi^i}(t, x(t)) = 0 \quad \text{für} \quad i = 2, \ldots, 6\,.$$

Ausgeschrieben lauten diese Bedingungen

$$y^2_{1t^2} + 2x^1\,|\sin t^2| - y_3^2 + \nu(t) - \mu(t) = 0\,, \tag{45}$$

$$\Phi_{t^1}x^2 + \psi_{2t^2}x^2 + 2x^2\,|\sin t^2| + \Phi x^6 + \psi_2 x^4 = 0\,,$$

$$y^2_{3t^2} - x^3\,|\sin t^2| + y_1^2 = 0\,, \qquad -x^4|\sin t^2| + \psi_2 x^2 = 0\,,$$

$$\frac{-x^5}{|\sin t^2|} = 0\,, \qquad \frac{-x^6}{|\sin t^2|} + \Phi x^2 = 0\,.$$

Da (x, u) ein optimaler und damit zulässiger Prozeß zu (40) sein soll, muß einerseits (40b–f) gelten und nach (5a) und (42) $u^1(t) = 0$ falls $y_3^2(t) < 0$ ist.

Unter Beachtung von (45) und der vorangehenden Bemerkung können wir die Bedingung (44) umformen zu

$$\Lambda(t, \xi) = a^2_{t^2} + (x^1)^2\,|\sin t^2| - (x^3)^2\,\frac{|\sin t^2|}{2} + (\xi^1 - x^1)\,(-\nu(t) + \mu(t)) + (\xi^1 - x^1)^2\left[\frac{\psi_{1t^2}}{2} + |\sin t^2|\right] -$$

$$- \frac{(\xi^3 - x^3)^2}{2}\,|\sin t^2| + \psi_1(\xi^1 - x^1)\,(\xi^3 - x^3) - \frac{(\xi^4)}{2}|\sin t^2| + (\xi^2)^2\left[\frac{\psi_{2t^2}}{2} + |\sin t^2|\right] +$$

$$+ \psi_2\xi^2\xi^4 - \frac{(\xi^5)^2 + (\xi^6)^2}{2\,|\sin t^2|} + \frac{\Phi_{t^1}}{2}\,(\xi^2)^2 + \Phi\xi^2\xi^6 \leqq 0 \tag{46}$$

für $t \in A_1 \cup A_2$ und $\Delta \leqq \xi^1 \leqq D$. Dabei ist $\nu(t)\,x^1(t) = \nu(t)\,\Delta$, $\mu(t)\,x^1(t) = \mu(t)\,D$.

Zum weiteren Studium von (46) halten wir vorerst $\xi^1, \ldots, \xi^5$ fest, dagegen ξ^6 variabel. Dann wird das Maximum von Λ bzgl. ξ^6 genau dort angenommen, wo $\Lambda_{\xi^6} = 0$ gilt, also für $\xi^6 = \Phi|\sin t^2|\,\xi^2$. Infolgedessen genügt es im weiteren, die Forderung (46) unter der Nebenbedingung $\xi^6 = \xi^2\Phi|\sin t^2|$ zu studieren bzw. nach entsprechender Elimination von ξ^6. Λ bildet dann ein Polynom zweiten Grades in $\xi^1, \ldots, \xi^5$ der Gestalt

$$\Lambda(t, \xi) = a^2_{t^2} + (x^1)^2\,|\sin t^2| - (x^3)^2\,\frac{|\sin t^2|}{2} + (\xi^1 - x^1)\,(-\nu(t) + \mu(t)) -$$

$$- \frac{(\xi^5)^2}{2|\sin t^2|} + Q_1 + Q_2\,, \tag{47}$$

wobei

$$Q_1 = \left(\frac{\psi_{1t^2}}{2} + |\sin t^2|\right)(\xi^1 - x^1)^2 + \psi_1(\xi^1 - x^1)\,(\xi^3 - x^3) - \frac{(\xi^3 - x^3)}{2}\,|\sin t^2| \tag{48a}$$

und

$$Q_2 = \left(\frac{\psi_{2t^2}}{2} + |\sin t^2| + \frac{\Phi^2}{2}|\sin t^2| + \frac{\Phi_{t^1}}{2}\right)(\xi^2)^2 + \psi_2\xi^2\xi^4 - \frac{\xi^4}{2}\cdot|\sin t^2| \tag{48b}$$

bedeuten.

Q_1 ist negativ semidefinit, falls

$$\psi_{1t^2}|\sin t^2| + \psi_1^2 + 2|\sin t^2|^2 = 0 \tag{49}$$

gilt. Für $x^1(t) = D$ und $\psi_1(t) = 0$ nimmt $Q_1 + \mu(t)\,(\xi^1 - D)$ unter der Nebenbedingung $\xi^1 \leqq D$ seinen Maximalwert für $\xi^1 = D$, $\xi^3 = x^3$ an.

Q_2 ist negativ semidefinit, falls

$$\psi_{2t^2}|\sin t^2| + 2|\sin t^2|^2 + \Phi^2|\sin t^2|^2 + \Phi_{t^1}|\sin t^2| + \psi_2^2 \leqq 0 \tag{50}$$

gilt.

Auf diesen Überlegungen aufbauend läßt sich nach einigen hier unterdrückten Rechnungen leicht das folgende **Resultat** zeigen:

Die Bedingungen (45), (46), (40 b–f) *und* $u^1(t)\, y_3^2(t) = 0$ *werden erfüllt durch das Funktionensystem*

$$x^1(t) = \begin{cases} D\cos(|t^2| - \alpha) & \text{für } |t^2| \leqq \alpha := \operatorname{arc\,cos}(\Delta/D) \\ D & \text{für } \alpha \leqq |t^2| \leqq \pi/2 \end{cases} \tag{51a}$$

$$u^1(t) = x^1_{t^2t^2} + x^1\,, \qquad x^3 = x^1_t\,, \qquad x^2 = x^4 = x^5 = x^6 \equiv 0;$$

$$\nu \equiv 0\,, \qquad \mu(t) = \begin{cases} 0 & \text{für } |t^2| \leqq \alpha\,, \\ 2D\sin|t^2| & \text{für } \alpha < |t^2| \leqq \pi/2; \end{cases} \tag{51b}$$

$$y_3^2(t) = \begin{cases} D\sin\alpha\left[1 - \cos t^2 - \left(\dfrac{1-\cos\alpha}{\sin\alpha}\right)\sin|t^2|\right] & \text{für } |t^2| \leqq \alpha\,, \\ 0 & \text{für } \alpha \leqq t^2 \leqq \pi/2\,, \end{cases} \tag{51c}$$

$$y_1^2(t) = \begin{cases} -D\sin((|t^2|) - \alpha)\sin|t^2| - D\sin\alpha\left[\sin|t^2| - \left(\dfrac{1-\cos\alpha}{\sin\alpha}\right)\cos t^2\right] & \text{für } |t^2| \leqq \alpha \\ 0 & \text{für } \alpha < |t^2| \leqq \pi/2; \end{cases} \tag{51d}$$

$$\psi_1(t) = \begin{cases} -\sin^2 t^2/\cos t^2 & \text{für } |t^2| < \alpha\,, \\ 0 & \text{für } \alpha < |t^2| \leqq \pi/2\,, \end{cases} \tag{51e}$$

$$\psi_2(t) = \begin{cases} -\sin^2 t^2/\cos t^2 & \text{für } |t^2| \leqq \alpha\,, \\ \varkappa\left(|t^2| - \dfrac{\pi}{2}\right)\Big/\left(\alpha - \dfrac{\pi}{2}\right) & \text{für } \alpha \leqq |t^2| \leqq \pi/2 \end{cases} \tag{51f}$$

mit $\varkappa = +\sin^2\alpha/\cos\alpha$;

$$\Phi(t) = -|B(t^2)|\tan(t^1\sin|t^2|) \quad \text{auf} \quad A_1 \cup A_2 \tag{51g}$$

mit $B^2(t^2) = \varkappa\sin|t^2| + \varkappa^2\left(|t^2| - \dfrac{\pi}{2}\right)^2 + 2\sin^2|t^2|$ (vgl. dazu auch Abb. 7).

a^2 berechnen wir aus der Bedingung $\Lambda(t, x(t)) = 0$ unter Integration über t^2. Die Bedingungen (5a), (5b) und (44) haben wir damit befriedigt.

Schließlich bestätigen wir noch das Erfülltsein der Bedingung (5c) und damit endgültig nach Satz 2 die Optimalität des Prozesses (x, u) bzgl. Problem (40).

Nach Definition (3) ist (unter Beachtung der Unstetigkeiten von ψ_1)

$$\hat{L}(S) = \inf_{\zeta\in\hat{Q}} \left\{\sum_{j=1}^{4} \int_{\partial\Omega_j} S^\alpha(t, \zeta(t))\, n^j_\alpha(t)\, do\right\}. \tag{52}$$

Dabei ist $\hat{Q}$ die Menge aller stetigen Vektorfunktionen $\zeta = (\zeta^1, \dots, \zeta^6)$ auf $A_1 \cup A_2$, die den Bedingungen (40c), (40e) und (40f) genügen und

$$\Omega_1 = \left[-\frac{\pi}{2}, \frac{\pi}{2}\right] \times \left[-\frac{\pi}{2}, -\alpha\right], \qquad \Omega_2 = \left[-\frac{\pi}{2}, \frac{\pi}{2}\right] \times [-\alpha, 0]\,,$$

$$\Omega_3 = \left[-\frac{\pi}{2}, \frac{\pi}{2}\right] \times [0, \alpha]\,, \qquad \Omega_4 = \left[-\frac{\pi}{2}, \frac{\pi}{2}\right] \times \left[\alpha, \frac{\pi}{2}\right].$$

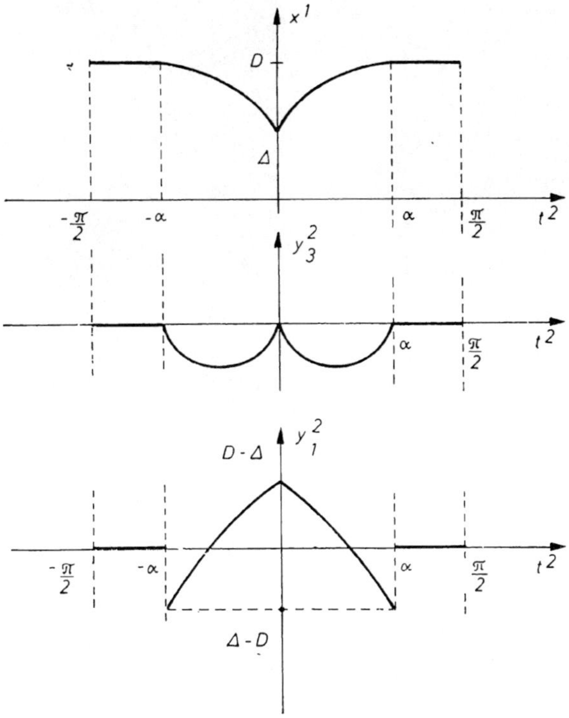

Abb. 7a

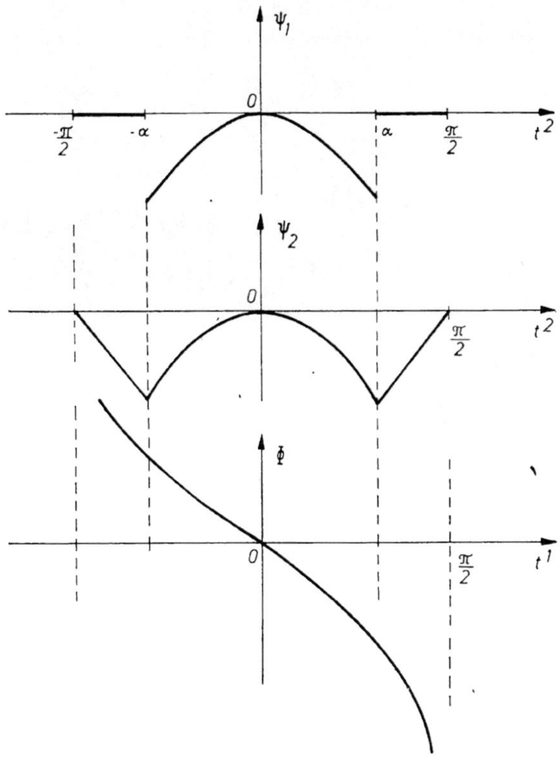

Abb. 7b

Somit ist

$$\hat{L}(S) = \inf_{\zeta \in \hat{Q}} \left\{ - \int_{-\pi/2}^{\pi/2} S^2\left(t^1, -\frac{\pi}{2}, \zeta\right)\Big|_{\Omega_1} \mathrm{d}t^1 + \int_{-\pi/2}^{\pi/2} S^2(t^1, -\alpha, \zeta)\Big|_{\Omega_1} \mathrm{d}t^1 - \int_{-\pi/2}^{\pi/2} S^2(t^1, -\alpha, \zeta)|_{\Omega_2} \mathrm{d}t^1 + \right.$$

$$+ \int_{-\pi/2}^{\pi/2} S^2(t^1, 0, \zeta)|_{\Omega_2} \mathrm{d}t^1 - \int_{-\pi/2}^{\pi/2} S^2(t^1, 0, \zeta)|_{\Omega_3} \mathrm{d}t^1 + \int_{-\pi/2}^{\pi/2} S^2(t^1, \alpha, \zeta)|_{\Omega_3} \mathrm{d}t^1 - \int_{-\pi/2}^{\pi/2} S^2(t^1, \alpha, \zeta)|_{\Omega_4} \mathrm{d}t^1 +$$

$$\left. + \int_{-\pi/2}^{\pi/2} S^2\left(t^1, \frac{\pi}{2}, \zeta\right)\Big|_{\Omega_4} \mathrm{d}t^1 + \lim_{\varepsilon \downarrow 0} \left[\int_{-\pi/2}^{\pi/2} S^1\left(\frac{\pi}{2} - \varepsilon, t^2, \zeta\right) \mathrm{d}t^2 - \int_{-\pi/2}^{\pi/2} S^1\left(-\frac{\pi}{2} + \varepsilon, t^2, \zeta\right) \mathrm{d}t^2 \right] \right\}. \quad (53)$$

Der Limes dieses Ausdrucks (53) erweist sich dabei als Null infolge (43a), (51g) und der Randbedingung (40f) (genauer besehen muß man dabei ähnliche Überlegungen führen wie in Bemerkung 2 zu Beispiel c)). Die Summe der restlichen Klammerglieder von (53) nimmt wegen (43b), (51) und (40e) ihr Minimum für $\zeta = x$ an. Damit erfüllen das konstruierte S und der Prozeß (x, u) gemäß (43)/(51) vollständig die Bedingung (5) und (2), so daß nach Satz 2 der Prozeß (x, u) optimal bzgl. des Steuerungsproblems (40) ist. Unsere gefundene optimale Zustandsfunktion x hat aber die Eigenschaft, über die Formeln (38) eine Funktion h zu definieren, die sogar sämtliche Nebenbedingungen von (13) erfüllt. Dieses zugeordnete h ist somit eine (nicht durchweg reguläre) Stützfunktion eines konvexen Bereichs $\mathfrak{B}$ aus $\mathfrak{A}_3(D, \Delta)$ mit der Eigenschaft $\operatorname{Min} J = J(x, u) = -O(h) = -\sup_{\mathfrak{A}_3(D,\Delta)} O$. Somit ist $\mathfrak{B}$ optimal für unser Problem d). Die Struktur dieses optimalen Bereichs wird durch folgendes zusammenfassendes Resultat beschrieben.

Behauptung 4: *Eine optimale Lösung des Problems* d) *ist ein Bereich* $\mathfrak{B}$, *der im* E^3 *als Durchschnitt eines Parallelstreifens* Σ *der Dicke* Δ *mit einer Kugel* $\mathfrak{B}_0$ *des Durchmessers* D *entsteht, wenn der Mittelpunkt* M *von* $\mathfrak{B}_0$ *von den Begrenzungsebenen zu* Σ *den gleichen Abstand besitzt.*

Literatur

1 Andrejewa, J. A.; Klötzler, R., Zur analytischen Lösung geometrischer Optimierungsaufgaben mittels Dualität bei Steuerungsproblemen. Teil I, ZAMM **64** (1984), 35–44.
2 Bonnesen, T.; Fenchel, W., Theorie der konvexen Körper, Berlin 1934.
3 Jaglom, J. M.; Boltjanski, W. G., Konvexe Figuren, Berlin 1956.
4 Kubota, T., Einige Ungleichheitsbeziehungen über Eilinien und Eiflächen, Sci. Rep. Tohoku Univ. **12** (1923), 45–65.

Eingegangen am 6. 12. 1982

Anschriften: Kand. d. Wiss. Jelena Arkadewna Andrejewa, Staatliche Universität Kalinin, Lehrstuhl für Numerische Mathematik, 170043 Kalinin/UdSSR; Prof. Dr. Rolf Klötzler, Karl-Marx-Universität Leipzig, Sektion Mathematik, Karl-Marx-Platz, DDR-7010 Leipzig

ZAMM 64, 155 – 162 (1984)

I. Y. Chen / G. M. Karadi / R. Y. Lai

Surface Selective Withdrawal in Sedimentation Basins

Mit Hilfe der Ausbreitungstheorie von blockförmigen Störungen werden allgemeine Bedingungen für die selektive Entnahme aus Aufschwemmbecken hergeleitet. Die Flüssigkeit bestehe aus zwei Schichten mit verschiedenen Dichteverteilungen. Der Einfluß einer freien Oberfläche wird in die Untersuchung einbezogen; die Zähigkeit wird, wie meist bei solchen Betrachtungen geschichteter Strömungen, unberücksichtigt gelassen. Es werden Kriterien für verschiedene Dichteprofile in Abhängigkeit von der densimetrischen Froude-Zahl aufgestellt. Eine selektive Entnahmeströmung wird vorkommen für densimetrische Froude-Zahlen, die kleiner sind als ihr kritischer Wert. Ein Vergleich der theoretischen Ergebnisse mit experimentellen Meßdaten von Bohan und Grace aus dem Jahre 1973 zeigt ausgezeichnete Übereinstimmung, solange die Dicke s der Bodenschicht klein ist verglichen mit der oberen. Die Ergebnisse der vorliegenden Arbeit wurden mit solchen verglichen, bei denen sich an Stelle der freien Oberfläche eine feste Grenze befindet. Für schwache Schichtung und kleine Werte von s zeigt sich ausgezeichnete Übereinstimmung. Wenn s wächst, wird für starke Schichtung der Einfluß der freien Oberfläche deutlich.

By employing the theory of propagation of columnar disturbance, general criteria for selective withdrawal in sedimentation basins are derived. The stratified fluid is assumed to be two-layered with different density distribution in each layer. The effects of a free surface are also included in the analysis, but the viscosity is neglected as in most other studies of stratified flow behavior. The criteria for different density profiles are established in terms of the densimetric Froude number. If the actual densimetric Froude number is smaller than its critical value, selective withdrawal will occur. Comparison of the theoretical results with the experimental data obtained by Bohan and Grace [1973] indicate an excellent agreement as long as the thickness of the bottom layer, s, is small compared to that of the top layer. Results obtained by assuming rigid upper boundary instead of free surface are also compared with the results derived in this paper. For weak stratification and small values of s they show excellent agreement. As s increases for strong stratification, the effect of free surface will become increasingly pronounced.

С помощью теории распространения столбчатых возмущений получены общие условия селективного изъятия из специального бассейна для седиментации. Предлогается, что жидкость состоит из двух слоев с разными распределениями плотностей. Включается в рассмотрение влияние свободной поверхности; вязкость, как это сделано во многих других исследованиях слоистых течений, не учтена. Найдены критерии для различных распределений плотности в зависимости от денсиметрического числа Фруда. Если денсиметрическое число Фруда меньше критической величины, то возникает селективное течение из бассейна. Сравнение теоретических результатов с экспериментальными данными Богана и Града 1973 года даёт отличное совпадение пока толщина s грунтового слоя мала по сравнению с толщиной верхнего слоя. Результаты данной работы также сравнивались с результатами, которые получились, если рассмотреть вместо свободной поверхности твёрдую границу. Для малой стратификации и малых параметров совпадение отличное. С ростом s для большой стратификации влияние свободной поверхности становится заметным.

Selective withdrawal of water from density stratified reservoirs has become a routine technique in water quality management. In most cases, it is practiced to receive water of a given temperature or of specific concentration of dissolved material and the withdrawal structures are located mainly near the bottom or at mid-level of the reservoir. For this reason theoretical studies of the hydrodynamics of selective withdrawal have been directed toward a better understanding of conditions governing bottom and mid-level withdrawal from reservoirs with density stratification caused by variation of temperature or dissolved material concentration. An extensive review of the literature on selective withdrawal can be found in the paper by Brooks and Koh [1969].

Background

Yih [1965] was probably the first investigator to propose a concise theoretical analysis of the selective withdrawal concept. A significant step toward a better understanding of its mechanism was achieved by Pao and Kao [1974]. Using perturbation theory for small values of the densimetric Froude number, they concluded that the phenomenon was controlled by columnar disturbance, i.e., by internal waves. Subsequently, Kao et al. [1964] proposed a criterion expressed in terms of the densimetric Froude number. According to the theory advanced by them, selective withdrawal will occur if the actual flow velocity is smaller than the velocity of the internal waves (disturbance) or, to put it in a different form, if the densimetric Froude number of the flow is below a critical value corresponding to the wave velocity.

In his study, Kao assumed a line sink located at the bottom. In order to analyze the error caused by replacing an opening of finite dimensions by a line sink, numerical studies have been performed by Pao and Kao [1974] and more recently by Lai [1979]. These studies concluded that the flow field is not sensitive to the size of the opening as long as it is small; hence, theoretical results obtained for line sink can be considered as good approximations in most practical situations.

As has been pointed out, most studies have analyzed bottom or mid-level selective withdrawal from reservoirs. The objective of this paper is to employ the selective withdrawal concept in the design of sedimentation basins. In a typical sedimentation basin, such as a wastewater settling basin and dredged material containment area, the depth is shallow and the density changes with depth due to concentration variation of suspended solid particles. With proper control of the water depth and the flowrate over the outlet weir, selective withdrawal may occur so that only the water near the surface with low solid concentration will flow over the weir and water at deeper layers with high solid concentration remains practically stagnant. Thus, the turbulent resuspension of bottom sediments can be prevented and, consequently, the effluent quality will be improved.

Most theoretical studies of the dynamics of stratified fluid involve two-dimensional flow bounded by two rigid plates. In this study, the effects of a free surface on the selective withdrawal criterion are analyzed. For the case of two-layer fluid, when the density difference is large and the depth of the withdrawal layer is small compared to that of the stagnant layer, the effect can be significant. As in most other studies, the viscosity of the mixture will be neglected. To investigate the effects of viscosity on the propagations of surface and internal waves in a two-layer fluid of finite depth, DIAZ [1980] used a matching expansion method. His results show that the viscous effect on the maximum speed of internal wave modes within the practical range of hydraulic parameters is insignificant, i.e., it has little influence on the selective withdrawal phenomenon.

Theoretical Analysis

A long sedimentation basin with a horizontal free surface and bottom is considered in this study. The fluid is assumed to be inviscid, incompressible and stratified with a typical mass density profile, shown in Fig. 1. The flowrate, Q, is induced by a line sink located at the top of the vertical wall downstream. For convenience, the origin of the coordinate system is selected at the interface of the flowing and stangant zones. The purpose of selective withdrawal is to let the fluid within the depth d_1 flow into the line sink while the lower layer of the fluid remains stagnant.

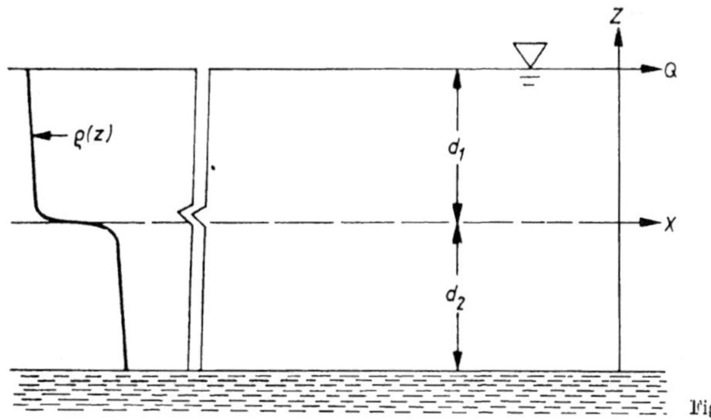

Fig. 1. Definition Sketch

Neglecting the convection acceleration terms in the Euler equations, PAO and KAO [1974] proposed the following partial differential equation governing two-dimensional flow in a stable stratified Boussinesq fluid:

$$\frac{\partial^2}{\partial t^2}\left(\frac{\partial^2 \psi}{\partial x^2}+\frac{\partial^2 \psi}{\partial z^2}\right)+N^2(z)\,\frac{\partial^2 \psi}{\partial x^2}=0\,, \tag{1}$$

in which the derivatives of the stream function ψ give the x and z components of velocity, $u = \partial\psi/\partial z$ and $v = -\partial\psi/\partial x$, respectively, and t denotes time. The coefficient $N(z)$ is the Brunt-Väisälä frequency defined as:

$$N^2 = -\frac{g}{\varrho_0}\frac{\mathrm{d}\varrho}{\mathrm{d}z}\,, \tag{2}$$

where ϱ_0 is a reference density and g is the gravitational acceleration.

For a specified density stratification, Eq. (1) is a partial differential equation with a variable coefficient. Since no general solution to Eq. (1) exists, it is necessary to introduce further simplification by considering N as a constant. Hence, it is assumed that the solution domain consists of discrete linearly stratified layers.

Using boundary layer approximation, i.e.,

$$\left|\frac{\partial^2}{\partial z^2}\right| \gg \left|\frac{\partial^2}{\partial x^2}\right|.$$

Eq. (1) reduces to

$$\frac{\partial^2}{\partial t^2}\left(\frac{\partial^2 \psi}{\partial z^2}\right)+N^2\,\frac{\partial^2 \psi}{\partial x^2}=0\,. \tag{3}$$

In this study, a propagation solution will be sought in the following form:

$$\psi(x, z, t) = \varphi(z)\, f(x + ct) \tag{4}$$

where c is the phase velocity of the propagating disturbance and φ denotes the amplitude. Substitution of Eq. (4) into Eq. (3) yields

$$\varphi'' + \frac{N^2}{c^2}\,\varphi = 0 \tag{5}$$

where φ'' is the second derivative of φ with respect to z.

In this study, the generalized stratification model shown in Fig. 2 has been selected as a good approximation of density distribution found in sedimentation basins. For this model:

$$\varrho(z) = \begin{cases} \varrho_{11} - (\varrho_{11} - \varrho_{21})(z/d_1), & 0 < z \leqq d_1, \\ \varrho_{21} - (\varrho_{20} - \varrho_{21})(z/d_2), & -d_2 \leqq z < d_1. \end{cases} \tag{6}$$

Hence, the corresponding Brunt-Väisälä frequency in each layer will be as follows:

$$\begin{cases} N_1^2 = -\dfrac{g}{\varrho_{11}}\dfrac{d\varrho}{dz} = \dfrac{g}{d_1}\left(\dfrac{\varrho_{11} - \varrho_{12}}{\varrho_{11}}\right), & 0 < z \leqq d_1, \\ N_2^2 = -\dfrac{g}{\varrho_{20}}\dfrac{d\varrho}{dz} = \dfrac{g}{d_2}\left(\dfrac{\varrho_{20} - \varrho_{21}}{\varrho_{20}}\right), & -d_2 \leqq z < d_1. \end{cases} \tag{7}$$

Substitution of N_1 and N_2 for N in Eq. (5) leads to

$$\begin{cases} \varphi_1'' + (N_1/c)^2 \varphi_1 = 0, & 0 < z \leqq d_1, \\ \varphi_2'' + (N_2/c)^2 \varphi_2 = 0, & -d_2 \leqq z < 0. \end{cases} \tag{8}$$

The eigenfunctions of φ_1 and φ_2 are:

$$\begin{cases} \varphi_1 = A_1 \sin\dfrac{N_1 z}{c} + A_2 \cos\dfrac{N_1 z}{c}, & 0 < z \leqq d_1, \\ \varphi_2 = A_3 \sin\dfrac{N_2 z}{c} + A_4 \cos\dfrac{N_2 z}{c}, & -d_2 \leqq z < 0, \end{cases} \tag{9}$$

in which A_1, A_2, A_3 and A_4 are integral constants to be determined from the boundary conditions.

At the bottom of the basin, the non-penetrating condition results in:

$$\varphi_2(-d_2) = 0. \tag{10}$$

At the interface ($z = 0$) between the upper and bottom layers, the pressure and the vertical velocity components are continuous, which lead to:

$$\varrho_{11}\varphi_1'(0) - \varrho_{21}\varphi_2'(0) + \frac{g}{c^2}[\varrho_{21}\varphi_2(0) - \varrho_{11}\varphi_1(0)] = 0 \tag{11}$$

and

$$\varphi_1(0) = \varphi_2(0). \tag{12}$$

(Derivation of Eq. (11) can be found, for instance, in the book by YIH [1980], pp. 25 and 26.)

The condition on the free surface is a special case of Eq. (11), when the "upper layer" (atmosphere) has a density of $\varrho = 0$. Hence, on the free surface ($z = d_1$)

$$\varphi_1'(d_1) = \frac{g}{c^2}\varphi_1(d_1). \tag{13}$$

Substitution of the boundary conditions given by Eqs. (10) through (13) into Eq. (9) leads to a system of homogeneous and linear algebraic equations:

$$\begin{aligned} &\left(\cos\frac{N_2 d_2}{c}\right) A_4 - \left(\sin\frac{N_2 d_2}{c}\right) A_3 = 0, \\ &\left(\frac{\varrho_{11} N_1}{c}\right) A_1 - \left(\frac{\varrho_{21} N_2}{c}\right) A_3 + \frac{g}{c^2}(g_{21} A_4 - g_{11} A_2) = 0, \\ &A_2 = A_4, \\ &\left(\frac{N_1}{c}\cos\frac{N_1 d_1}{c} - \frac{g}{c^2}\sin\frac{N_1 d_1}{c}\right) A_1 - \left(\frac{N_1}{c}\sin\frac{N_1 d_1}{c} - \frac{g}{c^2}\cos\frac{N_1 d_1}{c}\right) A_2 = 0. \end{aligned} \tag{14}$$

For non-trivial solutions to exist, the determinant of the coefficients has to be zero. By expanding the determinant and rearranging the terms,

$$\begin{aligned} &\sin\frac{N_2 d_2}{c}\left[\left(\frac{\varrho_{11}}{\varrho_{21}} - 1\right)\frac{g^2}{c^2}\sin\frac{N_1 d_1}{c} + \frac{g}{c} N_1 \cos\frac{N_1 d_1}{c} + \frac{\varrho_{11}}{\varrho_{21}} N_1^2 \sin\frac{N_1 d_1}{c}\right] - \\ &\quad - N_2 \cos\frac{N_2 d_2}{c}\left(N_1 \cos\frac{N_1 d_1}{c} - \frac{g}{c}\sin\frac{N_1 d_1}{c}\right) = 0 \end{aligned} \tag{15}$$

is obtained.

This dispersion equation can be satisfied by an infinite number of c values. The roots of Eq. (15), c_n, correspond to the phase velocity of various modes of the wave motion. For convenience, Eq. (15) will be expressed in dimensionless form:

$$\frac{\cos(rs\alpha)}{\sin(rs\alpha)} = \frac{\Delta\varrho_1(1 - \beta)\sin\alpha + \alpha\cos\alpha - (\beta\alpha^2\sin\alpha)/\Delta\varrho_1}{r(\Delta\varrho_1\cos\alpha - \alpha\sin\alpha)}, \tag{16}$$

in which $\alpha = N_1 d_1/c$, $r = N_2/N_1$, $s = d_2/d_1$, $\beta = 1 - \varrho_{11}/\varrho_{21}$, $\Delta\varrho_1 = 1 - \varrho_{21}/\varrho_{11}$.

It is interesting to point out that when $\varrho_{11} = \varrho_{21}$, i.e., there is no density discontinuity at the interface, the boundary condition given by Eq. (11) reduces to $\varphi_1'(0) = \varphi_2'(0)$. Hence, the horizontal component of the velocity is also continuous at the interface.

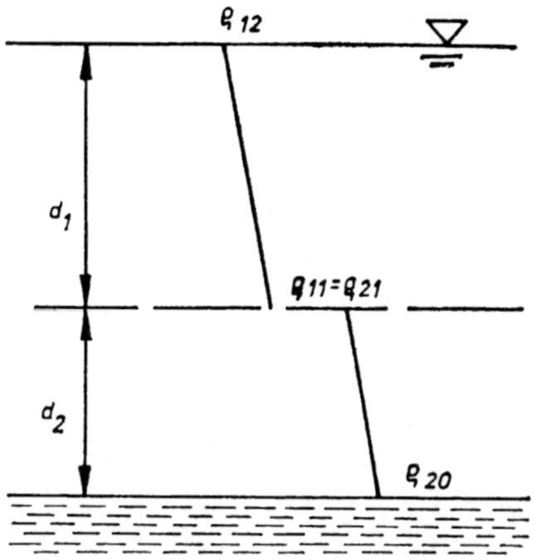

Fig. 2. Density Stratification Model

For the general case illustrated in Fig. 2, solution of Eq. (16) can be obtained by digital computer. In the special case of $N_1 = N_2 = 0$, i.e., when the density in each layer is constant, solution can be obtained directly. After substituting these values of N into Eq. (16) and rearranging the terms,

$$c^4 - g(d_1 + d_2)\, c^2 + \beta d_1 d_2 g^2 = 0 \tag{17}$$

is obtained for the calculation of the phase velocity. Solution of Eq. (17) for c^2 leads to:

$$c^2 = \frac{g}{2}\,[(d_1 + d_2) \pm \sqrt{(d_1 + d_2)^2 - 4\beta d_1 d_2}]\ . \tag{18}$$

The value of c^2 with a positive sign is the square of the speed of the surface waves. It was demonstrated by LAMB [1945] that there is only one mode of internal waves that exists in a two-layer model, and it is obtained from Eq. (18) if negative sign is used:

$$c_0 = \sqrt{\frac{g}{2}\,[(d_1 + d_2) - \sqrt{(d_1 + d_2)^2 - 4\beta d_1 d_2}]}\ . \tag{19}$$

For homogeneous fluid, $\beta = 0$, and $c = 0$, consequently, internal waves cannot develop.

According to the theory proposed by PAO and KAO [1974], selective withdrawal occurs if the columnar disturbance can travel upstream, i.e., if the phase velocity of the internal waves is greater than the initial flow velocity induced by sink. Hence, the speed of the fastest traveling mode of the internal waves has to be calculated from the dispersion relation to derive the appropriate selective withdrawal criterion. It should be pointed out that the root of Eq. (16), α_1, with the smallest positive value corresponds to the speed of surface waves, and α_2 having the second smallest positive value corresponds to the speed of the first mode of the internal waves. Thus, the criterion for selective withdrawal may be stated as follows:

$$\frac{q}{(d_1 + d_2)} = U < c_2 = \frac{N_1 d_1}{\alpha_2} \tag{20}$$

in which q is the discharge rate per unit width of the line sink, U is the initial uniform velocity of the flow and c_2 is the speed of the fastest traveling mode of the internal waves. If the densimetric Froude number, $F = U/N_1 d_1 = U/\sqrt{g d_1 \Delta\varrho_1}$ is smaller than its critical value

$$F_c = \frac{1}{\alpha_2}, \tag{21}$$

selective withdrawal will occur.

For the case of two homogeneous fluids, the densimetric Froude number may be expressed as $F = U/g d_1 \beta$. Hence, from Eq. (19) the critical Froude number of selective withdrawal can be expressed as:

$$F_c = \sqrt{\frac{1}{2\beta}\,[(1 + s) - \sqrt{(1 + s)^2 - 4\beta s}]}, \tag{22}$$

where $s = d_2/d_1$ is the depth ratio. When $\beta = 0$ or $s = 0$, i.e., for homogeneous fluid, it can be shown easily that $F_c = 0$, i.e., no selective withdrawal occurs as was shown earlier.

Parameter Study

For the purpose of determining the effect of density distribution and the relative dimensions of the upper and lower layers on selective withdrawal, a parameter study was performed. First, for the density stratification model shown in Fig. 2 the critical Froude number, F_c, was plotted against the density discontinuity parameter, $\beta = (\varrho_{21} - \varrho_{11})/\varrho_{21}$ for different depth ratios (Fig. 3). The results indicate that for $\beta < 0.01$ the critical Froude number is practically independent of β. However, as β increases, the effect of density discontinuity becomes more significant. The conclusion that F_c increases with β was expected, as greater β represents greater inertia of the stagnant layer. Fig. 3 also indicates that for a given density stratification, F_c increases with the depth ratio s. As s approaches zero, i.e., $d_2 \to 0$, the fluid becomes a single layer with linear density stratification and the value of F_c approaches $1/\pi$, which is the value predicted by YIH [1958] and later by PAO and KAO [1974].

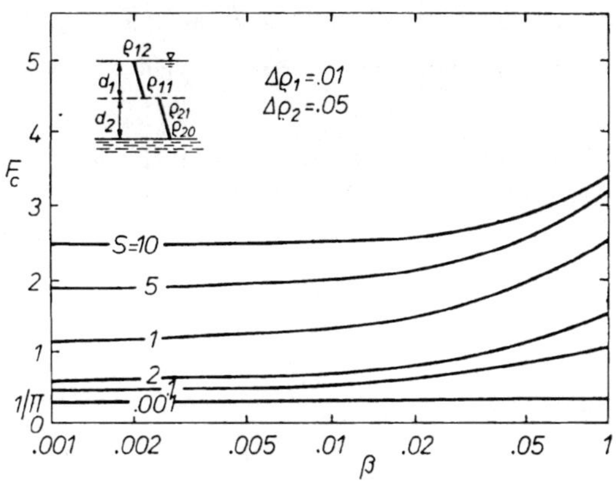

Fig. 3. Variation of Critical Froude Number with Density Discontinuity Parameter

For a basin with fixed total depth, $d_1 + d_2 =$ constant, it is useful to express the criterion of surface selective withdrawal in terms of q expressing the discharge rate per unit width of the basin. From Eqs. (20) and (21), the critical discharge rate for selective withdrawal is given by:

$$q_c = F_c(d_1 + d_2)\sqrt{g d_1 \Delta\varrho_1}\,. \tag{23}$$

Introducing a dimensionless discharge rate,

$$Q = \frac{q}{\sqrt{[(d_1 + d_2)^3\, g\, \Delta\varrho_1]}}, \tag{24}$$

the dimensionless critical discharge rate will be expressed by:

$$Q_c = \frac{F_c}{\sqrt{1 + s}}\,. \tag{25}$$

Taking the total depth constant, the variation of Q_c with depth ratio s for different values of $\Delta\varrho_1$, $\Delta\varrho_2$ and β is shown in Fig. 4. In this case the maximum critical discharge rate occurs near $s = 1$ for all densities, which suggests that for the purpose of effective operation of the basin it is desirable to withdraw water from the upper layer the thickness of which is one-half of the total depth.

If there is no discontinuity, $\beta = 0$. For this case, the variation of F_c with s for several values of $\Delta\varrho_1$ and $\Delta\varrho_2$ is presented in Fig. 5. For a given s, F_c increases with $\Delta\varrho_2$, but it decreases as $\Delta\varrho_1$ is becoming larger. For the same case, Fig. 6 shows the variation of Q_c with s. Again, the maximum critical discharge rate is found near $s = 1$.

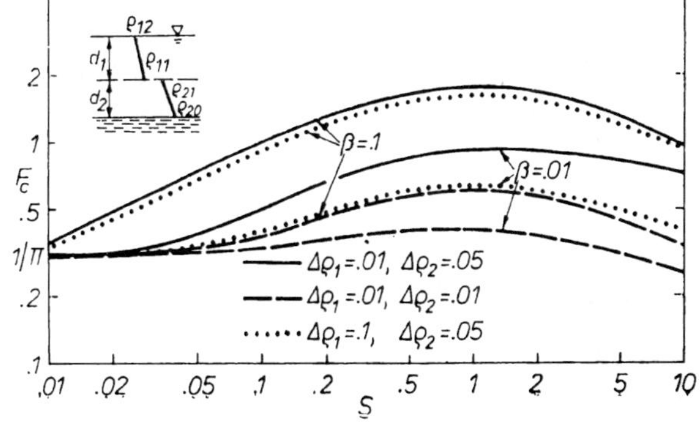

Fig. 4. Variation of Critical Discharge Rate with Depth Ratio for $\beta \neq 0$

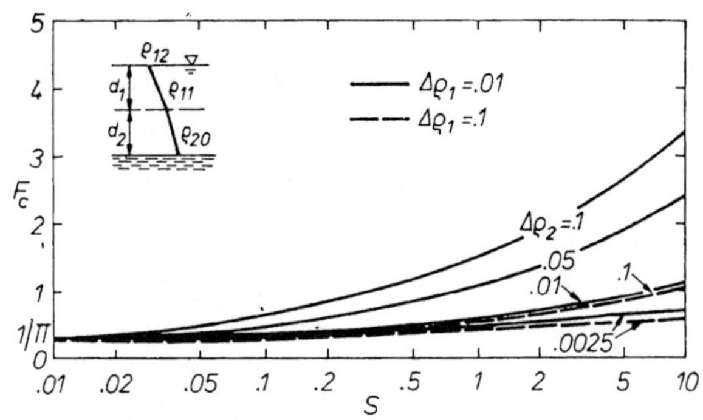

Fig. 5. Variation of Critical Froude Number with Depth Ratio for $\beta = 0$

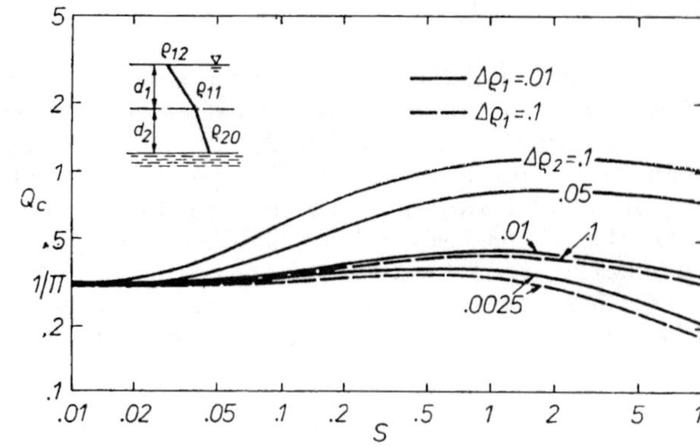

Fig. 6. Variation of Critical Discharge Rate with Depth Ratio for $\beta = 0$

Of special interest is the case when the density gradient of the withdrawal zone is the same as that of the stagnant zone, i.e., there is only one single layer with linear stratification. In this case the dispersion relation given by Eq. (16) reduces to

$$r = \frac{1}{\sqrt{1 + s\,\Delta\varrho_1}}.$$

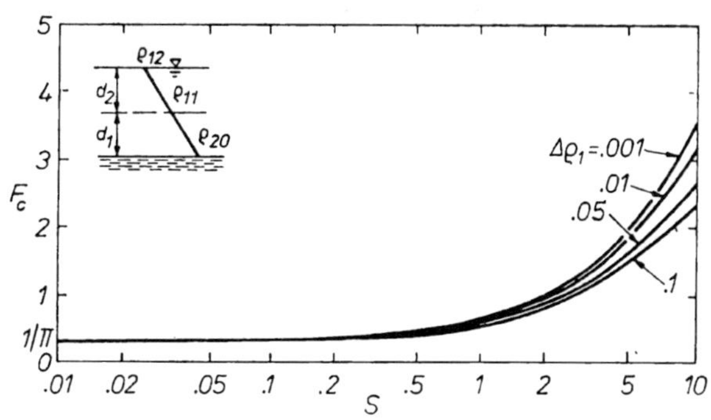

Fig. 7. Variation of Critical Froude Number with Depth Ratio for a Linearly Stratified Fluid

The variation of F_c with s for several values of $\Delta\varrho_1$ is shown in Fig. 7 for this special case.

The results of the theoretical model proposed herein have been compared with the experiments of BOHAN and GRACE [1973] who conducted model studies with stratified flow over a rectangular weir located in a channel of uniform width. Based on these studies, they proposed the following criterion for selective withdrawal:

$$F_T = \frac{q}{d_1 \sqrt{g d_1 \left(\dfrac{\Delta\varrho_W}{\varrho_W}\right)}} = 0.32\,, \tag{26}$$

in which F_T is a critical Froude number introduced by them, d_1 is the depth of the withdrawal zone, q is the discharge rate per unit width of the channel, ϱ_W is the fluid density at the level of the top of the weir, and $\Delta\varrho_W$ is the density difference between ϱ_W and the density at the bottom of the withdrawal zone. A definition sketch of their experiment

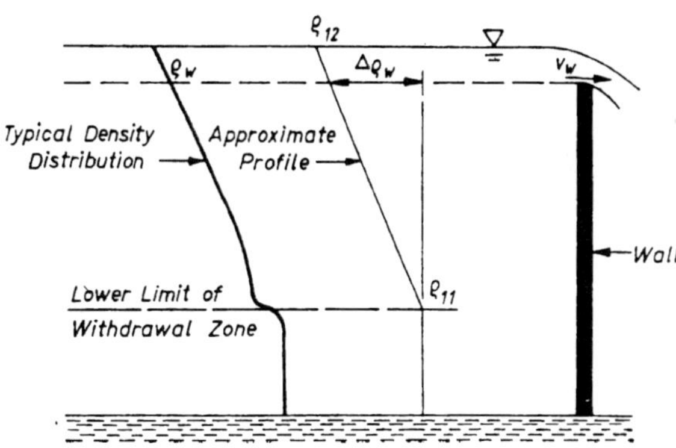

Fig. 8. Definition Sketch of Experimental Study by BOHAN and GRACE

is shown in Fig. 8. In Eq. (26) F_T is constant and independent of s and $\varrho(z)$. For the purpose of comparison, a typical density profile used in the experiments is approximated by a linear stratified layer overlaying a homogeneous layer ($N_2 = 0$). The relationship between F_T and F_c used in this study can be expressed as:

$$F_T = \sqrt{\frac{\varrho_{12}}{\varrho_{11}}}\, F_c\,. \tag{27}$$

In Fig. 9 the values of F_c are plotted against s for $\Delta\varrho_W = 0.01$ and 0.1. It is evident that for $s < 0.1$, the theoretical and experimental results are in excellent agreement. For this reason, and knowing that the values of s employed in the experiments of BOHAN and GRACE are small, it can be concluded that $F_T = 0.32$ is acceptable only if $d_2 \ll d_1$.

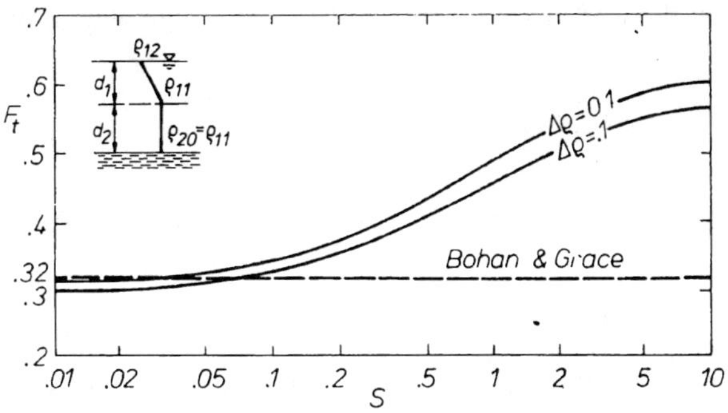

Fig. 9. Variation of Critical Froude Number with Depth Ratio for $\beta = 0$ and $N_2 = 0$

Most theoretical studies on the dynamics of stratified flow consider flow between two rigid boundaries. If the upper boundary of the flow is a free surface, assumption of a rigid surface instead will lead to erroneous results. The theoretical analysis of free surface flow is similar to that between two rigid plates except that the boundary condition at the upper boundary Eq. (11) is replaced by the non-penetrating condition, $\varphi_1(d_1) = 0$. For the purpose of demonstration, the two homogeneous fluid model will be considered. Using the dispersion equation advanced by LAMB [1945, p. 371], the speed of internal waves for two homogeneous fluids bounded by two horizontal plates can be expressed as:

$$c = \sqrt{\frac{g d_1 \beta s}{1 + (1 - \beta)\, s}}\,,$$

and the corresponding critical Froude number will be as follows:

$$F_{cr} = \frac{U}{\sqrt{g d_1 \beta}} = \sqrt{\frac{s}{1 + (1 - \beta)\, s}}\,. \tag{28}$$

In Fig. 10 the ratio of F_c/F_{cr}, (F_c is calculated from Eq. (22)), is plotted against s for different values of β. Since F_c is always smaller than F_{cr}, i.e., $F_c/F_{cr} < 1$, the speed of internal waves is faster if the surface is rigid. This may be explained by the fact that part of the energy, however small it may be, is converted into potential energy in the case of free surface. It is also clear from Fig. 10 that the difference between F_c and F_{cr} increases with both s and β. For small values of s and β, the difference is negligible, but for large values, it may become significant.

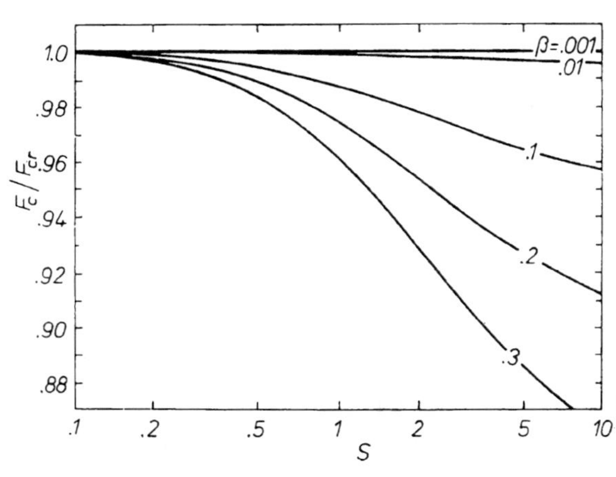

Fig. 10. Effects of Rigid Surface on Critical Froude Number

Summary and Conclusions

Employing the "columnar disturbance" theory, criteria of surface selective withdrawal in a two-dimensional stratified fluid have been developed. They are expressed in terms of functional relationships among the critical densimetric Froude number, F_c, the ratio of the depth of the stagnant zone to that of withdrawal zone, s, and the density profile, $\varrho(z)$. It has been found that F_c increases with s for all density profiles considered. The relation among the critical discharge rate, Q_c, s and $\varrho(z)$ for constant total depth was also studied. In this case the maximum critical discharge rate occurs at about $s = 1$. When comparing the theoretical results with the available experimental information, it can be concluded that within the practical range of data, the agreement is excellent. The theoretical results obtained for free surface have also been compared with the results derived for the case of rigid upper boundary. The difference appears to be negligible for small values of s and weak stratification; however, it may become significant for large values of s and for strong stratification. The results discussed in this paper can be applied to the operation and design of a sedimentation basin or dredged material disposal area for the control of the effluent quality.

References

1 BOHAN, J. P.; GRACE, J. L., Selective Withdrawal from Man-Made Lakes, Tech. Rep. H-73-4, Waterways Experiment Station, Vicksburg, Miss., March 1973.

2 BROOKS, N. H.; KOH, R. C. Y., Selective Withdrawal from Density-Stratified Reservoirs, Journal of the Hydraulics Division, ASCE, Vol. **95**, Proc. Paper 6702, July, No. HY4, 1969, pp. 1369–1400.

3 DIAZ, M., Selective Withdrawal in a Viscous Two-Layer Fluid, Thesis presented to the University of Wisconsin-Milwaukee, in May 1980, in partial fulfillment of the requirements for the degree of Master of Science.

4 KAO, T. W.; PAO, H. P.; WEI, S. N., Dynamics of Establishment of Selective Withdrawal from a Line Sink, Part II, Journal of Fluid Mechanics, Vol. **65**, 1974, pp. 689–710.

5 KAO, T. W., Selective Withdrawal Criteria of Stratified Fluids, Journal of the Hydraulics Division, ASCE, Vol. **102**, Proc. Paper 12188, June 1976, pp. 717–729.

6 LAI, R. Y., Finite Element Solution for Stratified Channel Flow, Proc. of the Specialty Conference on Conservation and Utilization of Water and Energy Resource, ASCE, S. F. Calif., August 1979, pp. 210–217.

7 LAMB, Sir H., Hydrodynamics, 6th ed., Dover Publications, New York, N.Y., 1945.

8 PAO, H. P.; KAO, T. W., Dynamics of Establishment of Selective Withdrawal from a Line Sink, Part I, Journal of Fluid Mechanics, Vol. **65**, 1974, pp. 657–688.

9 YIH, C. S., On the Flow of a Stratified Fluid, Proc. 3rd National Congress of Applied Mech., 1958, pp. 857–861.

10 YIH, C. S., Stratified Flows, Academic Press, New York, 1980.

Received 1983 March 23

Address: Dr. GABOR M. KARADI, Professor of Civil Engineering, University of Wisconsin, P.O. Box 784, Milwaukee, Wisconsin 53201, USA

ZAMM 64, 163 – 172 (1984)

Y. SUGANO

A General Treatise of Plane Thermoelastic Problem in Multiply-Connected Region Exhibiting Temperature-Dependent Properties

Ziel dieses Artikels ist es, die Formulierung des ebenen thermoelastischen Problems in einem mehrfach verbundenen Gebiet zu erweitern, indem die Temperaturabhängigkeiten der Materialeigenschaften in Termen der Airyschen Spannungsfunktion durch Ableitung der Bedingungen für die Eindeutigkeit der Drehung und der Verschiebungen, d. h. die modifizierten Michell-Bedingungen, berücksichtigt werden. Weiterhin soll eine Methode zur numerischen Anlyse des Systems der sich ergebenden Fundamentalgleichungen unter Benutzung der Finite-Differenzen-Methode vorgestellt werden.

The purpose of this paper is to complete the formulation of plane thermoelastic problem in a multiply-connected region taking into account the temperature dependencies of material properties in terms of Airy's stress function by deriving the conditions for the single valuedness of the rotation and displacements, i.e., modified Michell's conditions, and to present a method for analyzing the system of the resulting fundamental equations numerically by the use of finite difference method.

Целью работы является расширение формулировки плоской термоупругой задачи в многосвязной области учётом температурных зависимостей свойств материала посредством функции напряжения Эри с выводом условий единственности вращения и перемещений, т.е. модифицированных условий Мишеля. Кроме этого представляется метод для численного анализа основных уравнений путём использования метода конечных разностей.

1. Introduction

The determination of plane thermal stresses in multiply-connected region, i.e., finite region with many holes, plays an important role in the design of structures operating at elevated temperature such as nuclear reactors, supersonic aircrafts and gas and steam turbines.

In many papers [1–6] based on classical thermoelasticity, however, the solutions to such plane thermal-stress problems necessitating the consideration of the single-valuedness of the rotation and displacements have been obtained assuming that the material properties do not depend on the temperature, and do not truly describe the actual thermal-stress distributions in the structures mentioned above in which comparatively large temperature gradients occur. The studies on the thermal-stress problems in bodies having the temperature dependencies of material properties have been restricted to the one-dimensional problem [7, 8] even if the problems were solved by the finite difference method [9]. The analysis of the thermal stress intensity factor around a Griffith crack in a plate by HATA [10] seems to be the only work concerned with the thermal-stress problems in bodies having the temperature-dependent material properties under two-dimensional temperature field.

In this paper, therefore, we complete the formulation of plane thermoelastic problem in multiply-connected region in terms of Airy's stress function by deriving the conditions for the single-valuedness of rotation and displacements taking into account the temperature dependent material properties, i.e., modified Michell's conditions, and then present how the resulting fundamental equations are numerically solved by the use of finite difference method. Numerical results are obtained for a finite thin rectangular plate with a rectangular hole made of low-carbon steel.

2. Formulation of Plane Thermoelastic Problem in Multiply-Connected Region Exhibiting Temperature-Dependent Material Properties

Consider a general treatise of plane thermoelastic problem taking into account the temperature dependencies of material properties in a cylinder or plate in which the cross section is multiply connected. Let R be a connected region bounded by $n + 1$ non-intersecting contours $C_0, C_1, C_2, \ldots, C_n$ of which C_0 contains all the others as shown in Fig. 1. We use the Cartesian coordinates throughout the following formulation. Then the stress-strain relations are given by

$$\left.\begin{aligned} \varepsilon_{xx} &= [\sigma_{xx} - \nu_e(T)\,\sigma_{yy}]/E_e(T) + \Psi_e(T) \\ \varepsilon_{yy} &= [\sigma_{yy} - \nu_e(T)\,\sigma_{xx}]/E_e(T) + \Psi_e(T) \\ \varepsilon_{xy} &= [1 + \nu_e(T)]\,\sigma_{xy}/E_e(T) \end{aligned}\right\} \tag{1}$$

where

$$\left.\begin{aligned} &E_e(T) = E(T)\,, \qquad \nu_e(T) = \nu(T)\,, \qquad \Psi_e(T) = \int_0^T \alpha(\varrho)\,\mathrm{d}\varrho\,, \quad \text{for plane stress} \\ &E_e(T) = E(T)/[1 - \nu^2(T)]\,, \qquad \nu_e(T) = \nu(T)/[1 - \nu(T)]\,, \\ &\Psi_e(T) = [1 + \nu(T)] \int_0^T \alpha(\varrho)\,\mathrm{d}\varrho\,, \quad \text{for plane strain} \end{aligned}\right\} \tag{2}$$

and $E(T)$, $\nu(T)$ and $\alpha(T)$ are the Young's modulus, Poisson's ratio and linear thermal expansion coefficient, respectively.

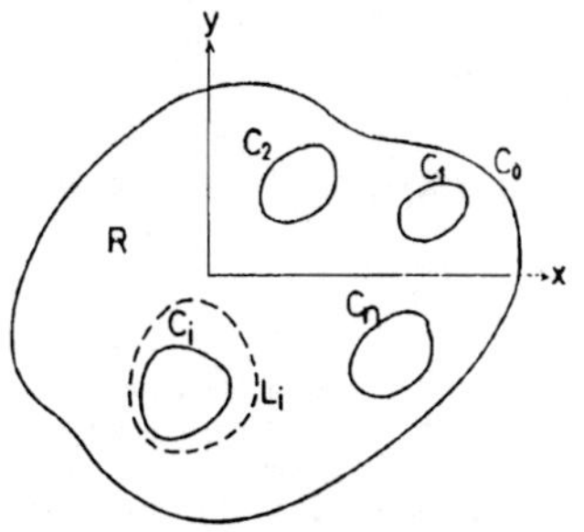

Fig. 1. Cross section of multiply-connected region considered in plane thermoelastic problem

The stresses in plane thermoelastic problem may be expressed in the following forms by Airy's stress function Φ defined in such a manner that the equations of equilibrium are identically satisfied:

$$\sigma_{xx} = \frac{\partial^2 \Phi}{\partial y^2}, \qquad \sigma_{yy} = \frac{\partial^2 \Phi}{\partial x^2}, \qquad \sigma_{xy} = -\frac{\partial^2 \Phi}{\partial x\, \partial y}. \tag{3}$$

The strain components must satisfy the compatibility equation

$$\frac{\partial^2 \varepsilon_{xx}}{\partial y^2} + \frac{\partial^2 \varepsilon_{yy}}{\partial x^2} = 2\frac{\partial^2 \varepsilon_{xy}}{\partial x\, \partial y}. \tag{4}$$

Substituting Eq. (1) into Eq. (4) and then expressing the stress components in terms of Airy's stress function by the use of Eq. (3), the fundamental differential equation for the plane thermoelastic problem taking into account the temperature dependencies of material properties are obtained as follows:

$$\begin{aligned}
&\nabla^4 \Phi - \frac{2}{E_e(T)}\left[\frac{\partial E_e(T)}{\partial x}\cdot\frac{\partial}{\partial x}(\nabla^2\Phi) + \frac{\partial E_e(T)}{\partial y}\cdot\frac{\partial}{\partial y}(\nabla^2\Phi)\right] + \\
&+ \left\{\frac{1}{E_e(T)}\left[\frac{2}{E_e(T)}\left(\frac{\partial E_e(T)}{\partial x}\right)^2 - \frac{\partial^2 E_e(T)}{\partial x^2}\right] - \frac{\nu_e(T)}{E_e(T)}\left[\frac{2}{E_e(T)}\left(\frac{\partial E_e(T)}{\partial y}\right)^2 - \frac{\partial^2 E_e(T)}{\partial y^2}\right] - \frac{\partial^2 \nu_e(T)}{\partial y^2} + \right. \\
&\left. + \frac{2}{E_e(T)}\frac{\partial E_e(T)}{\partial y}\frac{\partial \nu_e(T)}{\partial y}\right\}\frac{\partial^2\Phi}{\partial x^2} + \left\{\frac{1}{E_e(T)}\left[\frac{2}{E_e(T)}\left(\frac{\partial E_e(T)}{\partial y}\right)^2 - \frac{\partial^2 E_e(T)}{\partial y^2}\right] - \frac{\nu_e(T)}{E_e(T)}\left[\frac{2}{E_e(T)}\left(\frac{\partial E_e(T)}{\partial x}\right)^2 - \right.\right. \\
&\left.\left. - \frac{\partial^2 E_e(T)}{\partial x^2}\right] - \frac{\partial^2 \nu_e(T)}{\partial x^2} + \frac{2}{E_e(T)}\cdot\frac{\partial E_e(T)}{\partial x}\cdot\frac{\partial \nu_e(T)}{\partial x}\right\}\frac{\partial^2\Phi}{\partial y^2} + \\
&+ 2\left\{\frac{[1+\nu_e(T)]}{E_e(T)}\left[\frac{2}{E_e(T)}\frac{\partial E_e(T)}{\partial x}\frac{\partial E_e(T)}{\partial y} - \frac{\partial^2 E_e(T)}{\partial x\,\partial y}\right] + \frac{\partial^2 \nu_e(T)}{\partial x\,\partial y} - \frac{1}{E_e(T)}\frac{\partial E_e(T)}{\partial x}\cdot\frac{\partial \nu_e(T)}{\partial y} - \right. \\
&\left. - \frac{1}{E_e(T)}\cdot\frac{\partial E_e(T)}{\partial y}\cdot\frac{\partial \nu_e(T)}{\partial x}\right\}\frac{\partial^2\Phi}{\partial x\,\partial y} = -E_e(T)\,\nabla^2\Psi_e(T) \ldots
\end{aligned} \tag{5}$$

where

$$\nabla^2 = \frac{\partial^2}{\partial x^2} + \frac{\partial^2}{\partial y^2}.$$

Now the traction-free boundary conditions on C_i may be expressed in terms of Airy's stress function as follows [11]:

$$(\Phi)_{p_i} = c_{1i}x_{p_i} + c_{2i}y_{p_i} + c_{3i}, \tag{6}$$

$$\left(\frac{\partial \Phi}{\partial n'}\right)_{p_i} = c_{1i}\cos(x, n')_{p_i} + c_{2i}\cos(y, n')_{p_i} \tag{7}$$

where p_i is an arbitrary point on the i-th boundary C_i and n' is the arbitrary direction which does not coincide with the tangential direction to C_i. Since the addition of a linear function to Φ does not affect the stress distribution, one set of these constants c_{ji} ($j = 1, 2, 3$; $i = 0, 1, 2, \ldots, n$) in the multiply-connected region may be taken to be zero and it is usually convenient to take $c_{j0} = 0$, the conditions concerned with physical consideration must be used in order to determine the remaining $3n$ constants c_{jk} ($k = 1, 2, \ldots, n$). The conditions have been derived as the conditions for the single-valuedness of the rotation and displacements by MICHELL [12]. But there seems to be no work on the conditions which must be considered to ensure the single-valuedness of the rotation and displacements in the multiply-connected region exhibiting the temperature dependencies of material properties.

In this paper, therefore, we derived the conditions for the single-valuedness of the rotation and displacements taking into account the temperature dependencies of the material properties and expressed the resultings in terms of Airy's stress function as follows:

$$\oint_{L_i} \frac{\partial}{\partial n}\left[\frac{1}{E_e(T)}\nabla^2\Phi + \Psi_e(T)\right]\mathrm{d}s + \oint_{L_i}\frac{\partial}{\partial y}\left[\frac{1+\nu_e(T)}{E_e(T)}\right]d\left(\frac{\partial\Phi}{\partial x}\right) - \oint_{L_i}\frac{\partial}{\partial x}\left[\frac{1+\nu_e(T)}{E_e(T)}\right]\mathrm{d}\left(\frac{\partial\Phi}{\partial y}\right) = 0, \tag{8}$$

$$\oint_{L_i}\left(x\frac{\partial}{\partial s}-y\frac{\partial}{\partial n}\right)\left[\frac{1}{E_e(T)}\nabla^2\Phi+\Psi_e(T)\right]ds+\oint_{L_i}\left[\frac{1+\nu_e(T)}{E_e(T)}\right]d\left(\frac{\partial\Phi}{\partial x}\right)+$$
$$+\oint_{L_i}y\frac{\partial}{\partial x}\left[\frac{1+\nu_e(T)}{E_e(T)}\right]d\left(\frac{\partial\Phi}{\partial y}\right)-\oint_{L_i}y\frac{\partial}{\partial y}\left[\frac{1+\nu_e(T)}{E_e(T)}\right]d\left(\frac{\partial\Phi}{\partial x}\right)=0\,, \tag{9}$$

$$\oint_{L_i}\left(y\frac{\partial}{\partial s}+x\frac{\partial}{\partial n}\right)\left[\frac{1}{E_e(T)}\nabla^2\Phi+\Psi_e(T)\right]ds+\oint_{L_i}\left[\frac{1+\nu_e(T)}{E_e(T)}\right]d\left(\frac{\partial\Phi}{\partial y}\right)-$$
$$-\oint_{L_i}x\frac{\partial}{\partial x}\left[\frac{1+\nu_e(T)}{E_e(T)}\right]d\left(\frac{\partial\Phi}{\partial y}\right)+\oint_{L_i}x\frac{\partial}{\partial y}\left[\frac{1+\nu_e(T)}{E_e(T)}\right]d\left(\frac{\partial\Phi}{\partial x}\right)=0 \tag{10}$$

where L_i is a closed integral path including the boundary curve C_i in the interior of R, n is the normal direction to L_i and s represents the arc length along the curve L_i.

Finally, the formulation of the plane thermoelastic problem in the multiply-connected region exhibiting the temperature-dependent material properties in terms of Airy's stress function has been completed with Eqs. (5) to (10).

In general, there are no exact solutions to the plane thermoelastic problems in bodies under the two-dimensional temperature fields available by the use of the known analytical methods when the material properties depend on the temperature, or when the boundary curves are irregular, that is, not parallel to coordinates of a system. Hence we denote the next section to the finite difference representation of the resulting fundamental equations.

3. Finite Difference Representation of the Fundamental Equations

We take the finite difference grids with intervals Δx, Δy in the x and y directions, respectively and devide the multiply-connected region into a number of rectangular domains of small but finite size. Then the subscipt i is used to denote the i-th discrete point in the x direction and the subscript j is used to denote the j-th discrete point in the y direction.

The compatibility condition for the strain components, Eq. (4) may be expressed in the finite difference form by replacing the partial derivatives included in the equation by the equivalent central differences as follows:

$$A_1\tilde{\Phi}_{i,j}+A_2\tilde{\Phi}_{i+1,j}+A_3\tilde{\Phi}_{i-1,j}+A_4\tilde{\Phi}_{i,j+1}+A_5\tilde{\Phi}_{i,j-1}+A_6\tilde{\Phi}_{i+2,j}+A_7\tilde{\Phi}_{i-2,j}+A_8\tilde{\Phi}_{i,j+2}+$$
$$+A_9\tilde{\Phi}_{i,j-2}+A_{10}\tilde{\Phi}_{i+1,j+1}+A_{11}\tilde{\Phi}_{i+1,j-1}+A_{12}\tilde{\Phi}_{i-1,j+1}+A_{13}\tilde{\Phi}_{i-1,j-1}=-F(T_{i,j}) \tag{11}$$

where $\tilde{\Phi}=\Phi/l^2$, the following dimensionless quantities with certain reference length l were used

$$\xi=x/l\,,\qquad \eta=y/l\,,\qquad \Delta\xi=\Delta x/l\,,\qquad \Delta\eta=\Delta y/l\,, \tag{12}$$

and

$$\left.\begin{aligned}
A_1&=\frac{2}{E_{ei,j}}[3(\Delta\xi)^4+3(\Delta\eta)^4+4(\Delta\xi)^2(\Delta\eta)^2]-2[b_3(\Delta\xi)^4+b_2(\Delta\eta)^4+(b_1+b_4)(\Delta\xi)^2(\Delta\eta)^2]\,,\\
A_2&=-\frac{4(\Delta\eta)^2}{E_{ei,j}}[(\Delta\xi)^2+(\Delta\eta)^2]+(\Delta\eta)^2[(b_1-b_5)(\Delta\xi)^2+(b_2-b_5)(\Delta\eta)^2]\,,\\
A_3&=-\frac{4(\Delta\eta)^2}{E_{ei,j}}[(\Delta\xi)^2+(\Delta\eta)^2]+(\Delta\eta)^2[(b_1+b_5)(\Delta\xi)^2+(b_2+b_5)(\Delta\eta)^2]\,,\\
A_4&=-\frac{4(\Delta\xi)^2}{E_{ei,j}}[(\Delta\xi)^2+(\Delta\eta)^2]+(\Delta\xi)^2[(b_3-b_6)(\Delta\xi)^2+(b_4-b_6)(\Delta\eta)^2]\,,\\
A_5&=-\frac{4(\Delta\xi)^2}{E_{ei,j}}[(\Delta\xi)^2+(\Delta\eta)^2]+(\Delta\xi)^2[(b_3+b_6)(\Delta\xi)^2+(b_4+b_6)(\Delta\eta)^2]\,,\\
A_6&=\left(\frac{1}{E_{ei,j}}+\frac{b_5}{2}\right)(\Delta\eta)^4\,,\qquad A_7=\left(\frac{1}{E_{ei,j}}-\frac{b_5}{2}\right)(\Delta\eta)^4\,,\qquad A_8=\left(\frac{1}{E_{ei,j}}+\frac{b_6}{2}\right)(\Delta\xi)^4\,,\\
A_9&=\left(\frac{1}{E_{ei,j}}-\frac{b_6}{2}\right)(\Delta\xi)^4\,,\qquad A_{10}=\left[\frac{2}{E_{ei,j}}+\frac{1}{2}\left(b_5+b_6+\frac{b_7}{2}\right)\right](\Delta\xi)^2(\Delta\eta)^2\,,\\
A_{11}&=\left[\frac{1}{E_{ei,j}}+\frac{1}{2}\left(b_5-b_6-\frac{b_7}{2}\right)\right](\Delta\xi)^2(\Delta\eta)^2\,,\qquad A_{12}=\left[\frac{2}{E_{ei,j}}-\frac{1}{2}\left(b_5-b_6+\frac{b_7}{2}\right)\right](\Delta\xi)^2(\Delta\eta)^2\,,\\
A_{13}&=\left[\frac{2}{E_{ei,j}}-\frac{1}{2}\left(b_5+b_6-\frac{b_7}{2}\right)\right](\Delta\xi)^2(\Delta\eta)^2\,,\\
b_1&=-\nu_{ei,j}(E_{ei,j+1}-E_{ei,j-1})^2/(2E^3_{ei,j})+\\
&\quad+(E_{ei,j+1}-E_{ei,j-1})(\nu_{ei,j+1}-\nu_{ei,j-1})/(2E^2_{ei,j})+\\
&\quad+\nu_{ei,j}(E_{ei,j+1}+E_{ei,j-1}-2E_{ei,j})/E^2_{ei,j}-(\nu_{ei,j+1}+\nu_{ei,j-1}-2\nu_{ei,j})/E_{ei,j}\,,
\end{aligned}\right\} \tag{13}$$

$$\begin{aligned}
b_2 &= (E_{ei+1,j} - E_{ei-1,j})^2/(2E^3_{ei,j}) - (E_{ei+1,j} + E_{ei-1,j} - 2E_{ei,j})/E^2_{ei,j}\,, \\
b_3 &= (E_{ei,j+1} - E_{ei,j-1})^2/(2E^3_{ei,j}) - (E_{ei,j+1} + E_{ei,j-1} - 2E_{ei,j})/E^2_{ei,j}\,, \\
b_4 &= -\nu_{ei,j}(E_{ei+1,j} - E_{ei-1,j})^2/(2E^3_{ei,j}) + (E_{ei+1,j} - E_{ei-1,j})\,(\nu_{ei+1,j} - \nu_{ei-1,j})/(2E^2_{ei,i}) + \\
&\quad + \nu_{ei,j}(E_{ei+1,j} + E_{ei-1,j} - 2E_{ei,j})/E^2_{ei,j} - (\nu_{ei+1,j} + \nu_{ei-1,j} - 2\nu_{ei,j})/E_{ei,j}\,, \\
b_5 &= -(E_{ei+1,j} - E_{ei-1,j})/E^2_{ei,j}\,, \qquad b_6 = -(E_{ei,j+1} - E_{ei,j-1})/E^2_{ei,j}\,, \\
b_7 &= -(\nu_{ei+1,j} - \nu_{ei-1,j})\,(E_{ei,j+1} - E_{ei,j-1})/(2E^2_{ei,j}) + \\
&\quad + (1 + \nu_{ei,j})\,(E_{ei+1,j} - E_{ei-1,j})\,(E_{ei,j+1} - E_{ei,j-1})/E^3_{ei,j} - \\
&\quad - (1 + \nu_{ei,j})\,(E_{ei+1,j+1} - E_{ei+1,j-1} - E_{ei-1,j+1} + E_{ei-1,j-1})/(2E^2_{ei,j}) - \\
&\quad - (\nu_{ei,j+1} - \nu_{ei,j-1})\,(E_{ei+1,j} - E_{ei-1,j})/(2E^2_{ei,j}) + (\nu_{ei+1,j+1} - \nu_{ei+1,j-1} - \nu_{ei-1,j+1} + \\
&\quad + \nu_{ei-1,j-1})/(2E_{ei,j})\,,
\end{aligned}$$

$$\left.\begin{aligned}
F(T_{i,j}) &= (\Delta\xi)^2\,(\Delta\eta)^2\,[(\alpha_{i+1,j} - \alpha_{i-1,j})\,(T_{i+1,j} - T_{i-1,j})\,(\Delta\eta)^2/4 + \\
&\quad + (\alpha_{i,j+1} - \alpha_{i,j-1})\,(T_{i,j+1} - T_{i,j-1})\,(\Delta\xi)^2/4 + \alpha_{i,j}(T_{i+1,j} + \\
&\quad + T_{i-1,j} - 2T_{i,j})\,(\Delta\eta)^2 + \alpha_{i,j}(T_{i,j+1} + T_{i,j-1} - 2T_{i,j})\,(\Delta\xi)^2] \quad \text{for plane stress}\,, \\
F(T_{i,j}) &= (\Delta\xi)^2\,(\Delta\eta)^2 \Big\{[(\nu_{i+1,j} + \nu_{i-1,j} - 2\nu_{i,j})\,(\Delta\eta)^2 + (\nu_{i,j+1} + \nu_{i,j-1} - \\
&\quad - 2\nu_{i,j})\,(\Delta\xi)^2] \int_0^T \alpha(\varrho)\,\mathrm{d}\varrho + \frac{\alpha_{i,j}}{2}[(\nu_{i+1,j} - \nu_{i-1,j})\,(T_{i+1,j} - T_{i-1,j})\,(\Delta\eta)^2 + \\
&\quad + (\nu_{i,j+1} - \nu_{i,j-1})\,(T_{i,j+1} - T_{i,j-1})\,(\Delta\xi)^2] + \\
&\quad + \frac{(1 + \nu_{i,j})}{4}[(\alpha_{i+1,j} - \alpha_{i-1,j})\,(T_{i+1,j} - T_{i-1,j})\,(\Delta\eta)^2 + \\
&\quad + (\alpha_{i,j+1} - \alpha_{i,j-1})\,(T_{i,j+1} - T_{i,j-1})\,(\Delta\xi)^2] + (1 + \nu_{i,j})\,\alpha_{i,j} \times \\
&\quad \times [(T_{i+1,j} + T_{i-1,j} - 2T_{i,j})\,(\Delta\eta)^2 + (T_{i,j+1} + T_{i,j-1} - 2T_{i,j})\,(\Delta\xi)^2]\Big\} \\
&\qquad \text{for plane strain}\,.
\end{aligned}\right\} \tag{14}$$

The modified Michell's conditions, Eqs. (8)—(10), i.e., the conditions in order to ensure the single-valuedness of the rotation and displacements in the multiply-connected region with the temperature-dependent material properties may be expressed in the following finite difference form:

$$\oint_{L_i} (B_1\tilde{\Phi}_{i,j} + B_2\tilde{\Phi}_{i+1,j} + B_3\tilde{\Phi}_{i-1,j} + B_4\tilde{\Phi}_{i,j+1} + B_5\tilde{\Phi}_{i,j-1} + B_6\tilde{\Phi}_{i+2,j} + B_7\tilde{\Phi}_{i-2,j} + B_8\tilde{\Phi}_{i,j+2} + $$
$$+ B_9\tilde{\Phi}_{i,j-2} + B_{10}\tilde{\Phi}_{i+1,j+1} + B_{11}\tilde{\Phi}_{i+1,j-1} + B_{12}\tilde{\Phi}_{i-1,j+1} + B_{13}\tilde{\Phi}_{i-1,j-1} + B_{14})\,\mathrm{d}\bar{s} = 0\,, \tag{15}$$

$$\oint_{L_i} (D_1\tilde{\Phi}_{i,j} + D_2\tilde{\Phi}_{i+1,j} + D_3\tilde{\Phi}_{i-1,j} + D_4\tilde{\Phi}_{i,j+1} + D_5\tilde{\Phi}_{i,j-1} + D_6\tilde{\Phi}_{i+2,j} + D_7\tilde{\Phi}_{i-2,j} + D_8\tilde{\Phi}_{i,j+2} + $$
$$+ D_9\tilde{\Phi}_{i,j-2} + D_{10}\tilde{\Phi}_{i+1,j+1} + D_{11}\tilde{\Phi}_{i+1,j-1} + D_{12}\tilde{\Phi}_{i-1,j+1} + D_{13}\tilde{\Phi}_{i-1,j-1} + D_{14})\,\mathrm{d}\bar{s} = 0\,, \tag{16}$$

$$\oint_{L_i} (G_1\tilde{\Phi}_{i,j} + G_2\tilde{\Phi}_{i+1,j} + G_3\tilde{\Phi}_{i-1,j} + G_4\tilde{\Phi}_{i,j+1} + G_5\tilde{\Phi}_{i,j-1} + G_6\tilde{\Phi}_{i+2,j} + G_7\tilde{\Phi}_{i-2,j} + G_8\tilde{\Phi}_{i,j+2} + $$
$$G_9\tilde{\Phi}_{i,j-2} + G_{10}\tilde{\Phi}_{i+1,j+1} + G_{11}\tilde{\Phi}_{i+1,j-1} + G_{12}\tilde{\Phi}_{i-1,j+1} + G_{13}\tilde{\Phi}_{i-1,j-1} + G_{14})\,\mathrm{d}\bar{s} = 0 \tag{17}$$

where

$$\begin{aligned}
B_1 &= -2[(d_1 - d_2d_5)/(\Delta\xi)^2 + (d_1 - d_3d_4)/(\Delta\eta)^2]\,, \\
B_2 &= \frac{1}{\Delta\xi}\left\{\frac{(d_1 - d_2d_5)}{\Delta\xi} - \frac{d_4}{E_{ei,j}}\left[\frac{1}{(\Delta\xi)^2} + \frac{1}{(\Delta\eta)^2}\right]\right\}, \\
B_3 &= \frac{1}{\Delta\xi}\left\{\frac{(d_1 - d_2d_5)}{\Delta\xi} + \frac{d_4}{E_{ei,j}}\left[\frac{1}{(\Delta\xi)^2} + \frac{1}{(\Delta\eta)^2}\right]\right\}, \\
B_4 &= \frac{1}{\Delta\eta}\left\{\frac{(d_1 - d_3d_4)}{\Delta\eta} - \frac{d_5}{E_{ei,j}}\left[\frac{1}{(\Delta\xi)^2} + \frac{1}{(\Delta\eta)^2}\right]\right\}, \\
B_5 &= \frac{1}{\Delta\eta}\left\{\frac{(d_1 - d_3d_4)}{\Delta\eta} + \frac{d_5}{E_{ei,j}}\left[\frac{1}{(\Delta\xi)^2} + \frac{1}{(\Delta\eta)^2}\right]\right\}, \\
B_6 &= -B_7 = \frac{d_4}{2E_{ei,j}(\Delta\xi)^3}\,, \qquad B_8 = -B_9 = \frac{d_5}{2E_{ei,j}(\Delta\eta)^3}\,, \\
B_{10} &= \frac{1}{2\Delta\xi\,\Delta\eta}\left[\frac{(d_2d_4 + d_3d_5)}{2} + \frac{1}{E_{ei,j}}\left(\frac{d_5}{\Delta\xi} + \frac{d_4}{\Delta\eta}\right)\right],
\end{aligned}$$

$$
\left.
\begin{aligned}
B_{11} &= -\frac{1}{2\Delta\xi\,\Delta\eta}\left[\frac{(d_2 d_4 + d_3 d_5)}{2} + \frac{1}{E_{ei,j}}\left(\frac{d_5}{\Delta\xi} - \frac{d_4}{\Delta\eta}\right)\right], \\
B_{12} &= -\frac{1}{2\Delta\xi\,\Delta\eta}\left[\frac{(d_2 d_4 + d_3 d_5)}{2} - \frac{1}{E_{ei,j}}\left(\frac{d_5}{\Delta\xi} - \frac{d_4}{\Delta\eta}\right)\right], \\
B_{13} &= \frac{1}{2\Delta\xi\,\Delta\eta}\left[\frac{(d_2 d_4 + d_3 d_5)}{2} - \frac{1}{E_{ei,j}}\left(\frac{d_5}{\Delta\xi} + \frac{d_4}{\Delta\eta}\right)\right], \\
B_{14} &= \frac{\alpha_{i,j}}{2}\left[d_4(T_{i+1,j} - T_{i-1,j})/\Delta\xi + d_5(T_{i,j+1} - T_{i,j-1})/\Delta\eta\right] \quad \text{for plane stress}, \\
B_{14} &= \frac{1}{2}\left[\frac{d_4(\nu_{i+1,j} - \nu_{i-1,j})}{\Delta\xi} + \frac{d_5(\nu_{i,j+1} - \nu_{i,j-1})}{\Delta\eta}\right]\int_0^T \alpha(\varrho)\,\mathrm{d}\varrho + \\
&\quad + \frac{(1+\nu_{i,j})\,\alpha_{i,j}}{2}\left[\frac{d_4(T_{i+1,j} - T_{i-1,j})}{\Delta\xi} + \frac{d_5(T_{i,j+1} - T_{i,j-1})}{\Delta\eta}\right], \quad \text{for plane strain}, \\
d_1 &= -\frac{1}{2E_{ei,j}^2}\left[\frac{d_4(E_{ei+1,j} - E_{ei-1,j})}{\Delta\xi} + \frac{d_5(E_{ei,j+1} - E_{ei,j-1})}{\Delta\eta}\right], \\
d_2 &= -\frac{1}{2E_{ei,j}\Delta\eta}\left[\frac{(1+\nu_{i,j})}{E_{ei,j}}(E_{ei,j+1} - E_{ei,j-1}) - (\nu_{ei,j+1} - \nu_{ei,j-1})\right], \\
d_3 &= -\frac{1}{2E_{ei,j}\Delta\xi}\left[\frac{(1+\nu_{i,j})}{E_{ei,j}}(E_{ei+1,j} - E_{ei-1,j}) - (\nu_{ei+1,j} - \nu_{ei-1,j})\right], \\
d_4 &= \cos(\xi, \bar{n})_{i,j}, \qquad d_5 = \cos(\eta, \bar{n})_{i,j},
\end{aligned}
\right\} \tag{18}
$$

$$
\left.
\begin{aligned}
D_1 &= -2[(f_1 - f_4 f_5 + \eta d_2 d_5)/(\Delta\xi)^2 + (f_1 + \eta d_3 d_4)/(\Delta\eta)^2], \\
D_2 &= \frac{1}{\Delta\xi}\left\{\frac{(f_1 - f_4 f_5 + \eta d_2 d_5)}{\Delta\xi} - \frac{f_2}{E_{ei,j}}\left[\frac{1}{(\Delta\xi)^2} + \frac{1}{(\Delta\eta)^2}\right]\right\}, \\
D_3 &= \frac{1}{\Delta\xi}\left\{\frac{(f_1 - f_4 f_5 + \eta d_2 d_5)}{\Delta\xi} + \frac{f_2}{E_{ei,j}}\left[\frac{1}{(\Delta\xi)^2} + \frac{1}{(\Delta\eta)^2}\right]\right\}, \\
D_4 &= \frac{1}{\Delta\eta}\left\{\frac{(f_1 + \eta d_3 d_4)}{\Delta\eta} - \frac{f_3}{E_{ei,j}}\left[\frac{1}{(\Delta\xi)^2} + \frac{1}{(\Delta\eta)^2}\right]\right\}, \\
D_5 &= \frac{1}{\Delta\eta}\left\{\frac{(f_1 + \eta d_3 d_4)}{\Delta\eta} + \frac{f_3}{E_{ei,j}}\left[\frac{1}{(\Delta\xi)^2} + \frac{1}{(\Delta\eta)^2}\right]\right\}, \\
D_6 &= -D_7 = \frac{f_2}{2E_{ei,j}(\Delta\xi)^3}, \qquad D_8 = -D_9 = \frac{f_3}{2E_{ei,j}(\Delta\eta)^3}, \\
D_{10} &= \frac{1}{2\Delta\xi\,\Delta\eta}\left[\frac{(f_4 d_4 - \eta d_3 d_5 - \eta d_2 d_4)}{2} + \frac{1}{E_{ei,j}}\left(\frac{f_3}{\Delta\xi} + \frac{f_2}{\Delta\eta}\right)\right], \\
D_{11} &= -\frac{1}{2\Delta\xi\,\Delta\eta}\left[\frac{(f_4 d_4 - \eta d_3 d_5 - \eta d_2 d_4)}{2} + \frac{1}{E_{ei,j}}\left(\frac{f_3}{\Delta\xi} - \frac{f_2}{\Delta\eta}\right)\right], \\
D_{12} &= -\frac{1}{2\Delta\xi\,\Delta\eta}\left[\frac{(f_4 d_4 - \eta d_3 d_5 - \eta d_2 d_4)}{2} - \frac{1}{E_{ei,j}}\left(\frac{f_3}{\Delta\xi} - \frac{f_2}{\Delta\eta}\right)\right], \\
D_{13} &= \frac{1}{2\Delta\xi\,\Delta\eta}\left[\frac{(f_4 d_4 - \eta d_3 d_5 - \eta d_2 d_4)}{2} - \frac{1}{E_{ei,j}}\left(\frac{f_3}{\Delta\xi} + \frac{f_2}{\Delta\eta}\right)\right], \\
D_{14} &= \frac{\alpha_{i,j}}{2}[f_2(T_{i+1,j} - T_{i-1,j})/\Delta\xi + f_3(T_{i,j+1} - T_{i,j-1})/\Delta\eta] \quad \text{for plane stress}, \\
D_{14} &= \frac{1}{2}\left[\frac{f_2(\nu_{i+1,j} - \nu_{i-1,j})}{\Delta\xi} + \frac{f_3(\nu_{i,j+1} - \nu_{i,j-1})}{\Delta\eta}\right]\int_0^T \alpha(\varrho)\,\mathrm{d}\varrho + \\
&\quad + \frac{(1+\nu_{i,j})\,\alpha_{i,j}}{2}\left[\frac{f_2(T_{i+1,j} - T_{i-1,j})}{\Delta\xi} + \frac{f_3(T_{i,j+1} - T_{i,j-1})}{\Delta\eta}\right] \quad \text{for plane strain}, \\
f_1 &= -\frac{1}{2E_{ei,j}^2}\left[\frac{f_2(E_{ei+1,j} - E_{ei-1,j})}{\Delta\xi} + \frac{f_3(E_{ei,j+1} - E_{ei,j-1})}{\Delta\eta}\right], \\
f_2 &= -(\xi d_5 - \eta d_4)_{i,j}, \qquad f_3 = (\xi d_4 - \eta d_5)_{i,j}, \qquad f_4 = (1 + \nu_{ei,j})/E_{ei,j},
\end{aligned}
\right\} \tag{19}
$$

$$\left.\begin{aligned}
G_1 &= -2[(g_1 - \xi d_5)/(\Delta\xi)^2 + (g_1 + f_4 d_4 - \xi d_3 d_4)/(\Delta\eta)^2]\,,\\
G_2 &= \frac{1}{\Delta\xi}\left\{\frac{(g_1 - \xi d_5)}{\Delta\xi} - \frac{g_2}{E_{ei,j}}\left[\frac{1}{(\Delta\xi)^2} + \frac{1}{(\Delta\eta)^2}\right]\right\},\\
G_3 &= \frac{1}{\Delta\xi}\left\{\frac{(g_1 - \xi d_5)}{\Delta\xi} + \frac{g_2}{E_{ei,j}}\left[\frac{1}{(\Delta\xi)^2} + \frac{1}{(\Delta\eta)^2}\right]\right\},\\
G_4 &= \frac{1}{\Delta\eta}\left\{\frac{(g_1 + f_4 d_4 - \xi d_3 d_4)}{\Delta\eta} - \frac{g_3}{E_{ei,j}}\left[\frac{1}{(\Delta\xi)^2} + \frac{1}{(\Delta\eta)^2}\right]\right\},\\
G_5 &= \frac{1}{\Delta\eta}\left\{\frac{(g_1 + f_4 d_4 - \xi d_3 d_4)}{\Delta\eta} + \frac{g_3}{E_{ei,j}}\left[\frac{1}{(\Delta\xi)^2} + \frac{1}{(\Delta\eta)^2}\right]\right\},\\
G_6 &= -G_7 = \frac{g_2}{2E_{ei,j}(\Delta\xi)^2}\,, \qquad G_8 = -G_9 = \frac{g_3}{2E_{ei,j}(\Delta\eta)^2}\,,\\
G_{10} &= \frac{1}{2\Delta\xi\,\Delta\eta}\left[\frac{(-f_4 d_5 + \xi d_3 d_5 + \xi d_2 d_4)}{2} + \frac{1}{E_{ei,j}}\left(\frac{g_3}{\Delta\xi} + \frac{g_2}{\Delta\eta}\right)\right],\\
G_{11} &= -\frac{1}{2\Delta\xi\,\Delta\eta}\left[\frac{(-f_4 d_5 + \xi d_3 d_5 + \xi d_2 d_4)}{2} + \frac{1}{E_{ei,j}}\left(\frac{g_3}{\Delta\xi} - \frac{g_2}{\Delta\eta}\right)\right],\\
G_{12} &= -\frac{1}{2\Delta\xi\,\Delta\eta}\left[\frac{(-f_4 d_5 + \xi d_3 d_5 + \xi d_2 d_4)}{2} - \frac{1}{E_{ei,j}}\left(\frac{g_3}{\Delta\xi} - \frac{g_2}{\Delta\eta}\right)\right],\\
G_{13} &= \frac{1}{2\Delta\xi\Delta\eta}\left[\frac{(-f_4 d_5 + \xi d_3 d_5 + \xi d_2 d_4)}{2} - \frac{1}{E_{ei,j}}\left(\frac{g_3}{\Delta\xi} + \frac{g_2}{\Delta\eta}\right)\right],\\
G_{14} &= \frac{\alpha_{i,j}}{2}[g_2(T_{i+1,j} - T_{i-1,j})/\Delta\xi + g_3(T_{i,j+1} - T_{i,j-1})/\Delta\eta] \quad \text{for plane stress},\\
G_{14} &= \frac{1}{2}[g_2(\nu_{i+1,j} - \nu_{i-1,j})/\Delta\xi + g_3(\nu_{i,j+1} - \nu_{i,j-1})/\Delta\eta]\int_0^T \alpha(\varrho)\,d\varrho\, +\\
&\quad + \frac{(1 + \nu_{i,j})\,\alpha_{i,j}}{2}\left[\frac{g_2(T_{i+1,j} - T_{i-1,j})}{\Delta\xi} + \frac{g_3(T_{i,j+1} - T_{i,j-1})}{\Delta\eta}\right] \quad \text{for plane strain},\\
g_1 &= -\frac{1}{2E_{ei,j}^2}\left[\frac{g_2(E_{ei+1,j} - E_{ei-1,j})}{\Delta\xi} + \frac{g_3(E_{ei,j+1} - E_{ei,j-1})}{\Delta\eta}\right],\\
g_2 &= (\xi d_4 - \eta d_5)_{i,j}\,, \qquad g_3 = (\xi d_5 + \eta d_4)_{i,j}\,.
\end{aligned}\right\} \tag{20}$$

The integrations of Eqs. (15)—(17) are evaluated by means of the numerical integral method (i.e., Simpson's rule).

Finally, to solve by the finite difference method the plane thermoelastic problems in the multiply-connected region having the temperature-dependent material properties formulated in terms of Airy's stress function is reduced to the determination of the values of stress function at N nodal points in the interior of R and $3n$ unknown constants c_{1i}, c_{2i}, c_{3i} by solving the simultaneous equations composed of the applications of Eq. (11) to N nodal points and of Eqs. (15)—(17) to n closed integral paths (L_i, $i = 1, 2, \ldots, n$) including the boundary curves (C_i, $i = 1, 2, \ldots, n$). Then, the values of the stress function at the nodal points on the boundaries C_i are easily determined by the substitution of the constants c_{ji} ($j = 1, 2, 3$; $i = 1, 2, \ldots, n$) into Eq. (6).

4. Numerical Calculations and Discussion

Consider the transient temperature and plane thermal stress distributions in a thin rectangular plate of dimension $2l_x \times 2l_y$ and thickness b with a rectangular hole of dimension $2l'_x \times 2l'_y$ as a case of the multiply-connected region

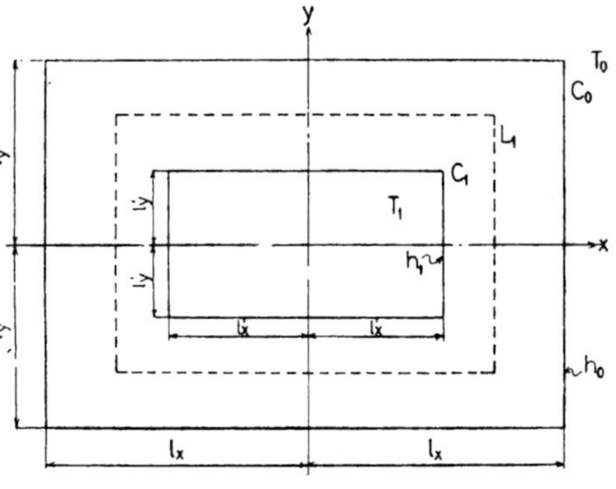

Fig. 2. Rectangular plate with a rectangular hole and a closed contour L_1

as shown in Fig. 2, assuming that the plate, initially at the same uniform temperature T_0 as the surrounding media, is heated by the abrupt change T_1 of the inner surrounding temperature and that there is a heat loss into the surrounding medium at constant temperature T_0 on the outer boundary and the upper and lower surfaces. Taking into account the temperature dependencies of the material properties and the heat loss from the upper and lower surfaces through the heat transfer coefficient h, we may express the governing equation for the transient temperature field in the plate as follows:

$$\frac{\partial}{\partial x}\left[\lambda(T)\frac{\partial T}{\partial x}\right]+\frac{\partial}{\partial y}\left[\lambda(T)\frac{\partial T}{\partial y}\right]-2h(T-T_0)/b=c(T)\,\gamma(T)\frac{\partial T}{\partial t}. \tag{21}$$

From the above description, the initial and boundary conditions may be expressed as

$$T=T_0\,,\qquad t=0\,, \tag{22}$$

$$\lambda(T)\frac{\partial T}{\partial x}+h_0(T-T_0)=0,\qquad x=l_x\,, \tag{23}$$

$$\lambda(T)\frac{\partial T}{\partial y}+h_0(T-T_0)=0\,,\qquad y=l_y\,, \tag{24}$$

$$\lambda(T)\frac{\partial T}{\partial x}-h_1(T-T_1)=0\,,\qquad x=l'_x\,, \tag{25}$$

$$\lambda(T)\frac{\partial T}{\partial y}-h_1(T-T_1)=0\,,\qquad y=l'_y\,. \tag{26}$$

We may obtain the following finite difference forms of Eqs. (21)—(26) by replacing the partial derivatives by the central differences in the spatial coordinates and the pure implicit method in time:

$$\begin{aligned}&[(\lambda^{*k}_{i+1,j}+2\lambda^{*k}_{i,j}+\lambda^{*k}_{i-1,j})/(\Delta\xi)^2+(\lambda^{*k}_{i,j+1}+2\lambda^{*k}_{i,j}+\lambda^{*k}_{i,j-1})/(\Delta\eta)^2+(\varrho^{*k}_{i,j}+\varrho^{*k-1}_{i,j})/\Delta\tau+4m/\beta]\,T^k_{i,j}-\\&-(\lambda^{*k}_{i+1,j}+\lambda^{*k}_{i,j})\,T^k_{i+1,j}/(\Delta\xi)^2-(\lambda^{*k}_{i,j}+\lambda^{*k}_{i-1,j})\,T^k_{i-1,j}/(\Delta\xi)^2-(\lambda^{*k}_{i,j+1}+\lambda^{*k}_{i,j})\,T^k_{i,j+1}/(\Delta\eta)^2-\\&-(\lambda^{*k}_{i,j}+\lambda^{*k}_{i,j-1})\,T^k_{i,j-1}/(\Delta\eta)^2=(\varrho^{*k}_{i,j}+\varrho^{*k-1}_{i,j})\,T^{k-1}_{i,j}/\Delta\tau+4mT_0/\beta\,,\end{aligned} \tag{27}$$

$$T^0_{i,j}=T_0\,,\qquad \tau=0\,, \tag{28}$$

$$\lambda^{*k}_{i,j}(T^k_{i+1,j}-T^k_{i-1,j})/(2\Delta\xi)+m_0(T^k_{i,j}-T_0)=0\,,\qquad \xi=1\,, \tag{29}$$

$$\lambda^{*k}_{i,j}(T^k_{i,j+1}-T^k_{i,j-1})/(2\Delta\eta)+m_0(T^k_{i,j}-T_0)=0\,,\qquad \eta=\bar{l}_y\,, \tag{30}$$

$$\lambda^{*k}_{i,j}(T^k_{i+1,j}-T^k_{i-1,j})/(2\Delta\xi)-m_1(T^k_{i,j}-T_1)=0\,,\qquad \xi=\bar{l}'_x\,, \tag{31}$$

$$\lambda^{*k}_{i,j}(T^k_{i,j+1}-T^k_{i,j-1})/(2\Delta\eta)-m_1(T^k_{i,j}-T_1)=0\,,\qquad \eta=\bar{l}'_y \tag{32}$$

where the thermal conductivity λ, the specific heat c and the density γ were disjointed into two factors, the first one dimensional and invariant, denoted by the subscript nought, and the second one dimensionless and a function of temperature, denoted by an asterisk:

$$\lambda(T)=\lambda_0\lambda^*(T)\,,\qquad c(T)=c_0c^*(T)\,,\qquad \gamma(T)=\gamma_0\gamma^*(T)\,, \tag{33}$$

and the following dimensionless quantities with the half-length of the plate l_x as a reference length were introduced:

$$\left.\begin{aligned}&\xi=x/l_x\,,\quad \eta=y/l_x\,,\quad \tau=\varkappa_0t/l_x^2\,,\quad \varkappa_0=\lambda_0/(c_0\gamma_0)\,,\quad m=hl_x/\lambda_0\,,\quad m_0=h_0l_x/\lambda_0\,,\\&m_1=h_1l_x/\lambda_0\,,\quad \bar{l}_y=l_y/l_x\,,\quad \bar{l}'_x=l'_x/l_x\,,\quad \bar{l}'_y=l'_y/l_x\,,\quad \varrho^*(T)=c^*(T)\,\gamma^*(T)\,.\end{aligned}\right\} \tag{34}$$

In Eqs. (27)—(32) the superscript k denotes the unknown values at the new dimensionless time $\tau=\tau_k$ and the superscript $k-1$ the known ones at $\tau=\tau_{k-1}$. The trouble in Eq. (27) is that the thermal properties $\lambda^{*k}_{i,j}$ and $\varrho^{*k}_{i,j}$ at $\tau=\tau_k$ depend upon the unknown temperatures at $\tau=\tau_k$ that we are looking for. We used an iterative procedure to solve this nonlinear heat-conduction problem. At the first step we use the thermal properties for temperatures at τ τ_{k-1} (i.e., $\lambda^{*k-1}_{i,j}$, $\varrho^{*k-1}_{i,j}$) as a first approximation of the thermal properties at $\tau=\tau_k$ in Eq. (27) and solve Eq. (27) for the unknown temperature $T^k_{i,j}$. At the second step we recompute the thermal properties at $\tau=\tau_k$ by the use of the temperatures that we have just founded, and then we obtain an improved solution for the temperatures $T^k_{i,j}$ at $\tau=\tau_k$ by solving again Eq. (27) for the temperatures. This process was continued until the sufficient convergence to the temperatures was obtained, and then we put the calculation at the next time step.

Thus, when the values of the temperature at the nodal points have been determined at each time step and the temperature dependencies of the material properties have been prescribed, we can evaluate the values of A_1-A_{13} and $F(T_{i,j})$ in the fundamental equation of the plane thermoelasticity and the values of B_1-B_{14}, D_1-D_{14}, and G_1-G_{14} in the modified Michell's conditions (15)—(17) which are represented in the finite difference forms. In the problem as shown in Fig. 2, the equations (16) and (17) necessary for the single-valuedness of the displacement components u_x and u_y are identically satisfied because both the temperature and thermal stress fields are symmetric with respect to x- and y-axes.

The mechanical boundary conditions in terms of Airy's stress function in this case may be expressed as follows:

$$\left.\begin{aligned}
&\tilde{\Phi}_{i,j} = \frac{\tilde{\Phi}_{i+1,j} - \tilde{\Phi}_{i-1,j}}{2\Delta\xi} = 0\,, && \xi = 1\,,\\
&\tilde{\Phi}_{i,j} = \frac{\tilde{\Phi}_{i,j+1} - \tilde{\Phi}_{i,j-1}}{2\Delta\eta} = 0\,, && \eta = \bar{l}_y\,,\\
&\tilde{\Phi}_{i,j} = \tilde{c}_{31}\,, \quad \tilde{c}_{31} = c_{31}/l_x^2\,, \quad \frac{\tilde{\Phi}_{i+1,j} - \tilde{\Phi}_{i-1,j}}{2\Delta\xi} = 0\,, && \xi = \bar{l}'_x\,,\\
&\tilde{\Phi}_{i,j} = \tilde{c}_{31}\,, \quad \frac{\tilde{\Phi}_{i,j+1} - \tilde{\Phi}_{i,j-1}}{2\Delta\eta} = 0\,, && \eta = \bar{l}'_y\,.
\end{aligned}\right\} \quad (35)$$

Finally, the plane thermal-stress problem in the rectangular plate with a rectangular hole exhibiting the temperature dependencies of the material properties is reduced to the determination of the values of stress function $\tilde{\Phi}_{i,j}$ at N nodal points in the interior of the region and the constant $\bar{c}_{31}$ by solving simultaneously the equations obtained from the application of Eq. (11) to N nodal points and Eq. (15) integrated numerically with the aid of Simpson's rule along the closed contour L_1 in Fig. 2. Thus, once the values of the stress function are determined at all nodal points including the nodal points on the inner and outer boundaries, the thermal-stress components can be easily calculated by the use of the finite difference representation of Eq. (3).

Numerical calculations were carried out for two cases of the rectangular plate with a rectangular hole made of low-carbon steel having the following material properties [13]—[15]:

$$\varrho(T) = c(T)\,\gamma(T) = 0.842 + 0.791 \times 10^{-3}T \text{ (cal/°C} \cdot \text{cm}^3) = 3.53 \times 10^3 + 3.31T \text{ (KJ/Kg} \cdot \text{K)}\,,$$
$$\lambda(T) = 0.112 - 0.614 \times 10^{-4}T \text{ (cal/cm} \cdot \text{s} \cdot \text{°C)} = 0.469 \times 10^{-1} - 0.257 \times 10^{-4}T \text{ (KW/m} \cdot \text{K)}\,,$$
$$E(T) = 0.220 \times 10^5 - 0.250 \times 10^{-1}T^2 \text{ (Kgf/mm}^2) = 0.216 \times 10^3 - 0.245 \times 10^{-3}T^2 \text{ (GPa)}\,,$$
$$\alpha(T) = 0.119 \times 10^{-4} + 0.832 \times 10^{-8}T \text{ (1/°C)},$$
$$\nu(T) = 0.3\,.$$

Case (1): $\bar{l}_y = 1$, $\bar{l}'_x$, $\bar{l}'_y = 0.5$, $m_0 = 1$, $m_1 = \infty$, $m = 0$, $T_0 = 100$ °C, $T_1 = 200$ °C, $\beta = 0.05$;

Case (2): $\bar{l}_y = 0.5$, $\bar{l}'_x = 0.5$, $\bar{l}'_y = 0.25$, $m_0 = 1$, $m_1 = \infty$, $m = 0$, $T_0 = 100$ °C, $T_1 = 200$ °C, $\beta = 0.05$.

The spatial and time intervals for all cases calculated were taken as $\Delta\xi = \Delta\eta = 0.03125$ and $\Delta\tau = 0.0025$, respectively.

In all figures from now on, the solid lines, the dashed-point lines and the dashed lines indicate the transient temperature or thermal stress distributions calculated by the use of the material constants exhibiting the temperature dependence and at the initial temperature and the room temperature, respectively.

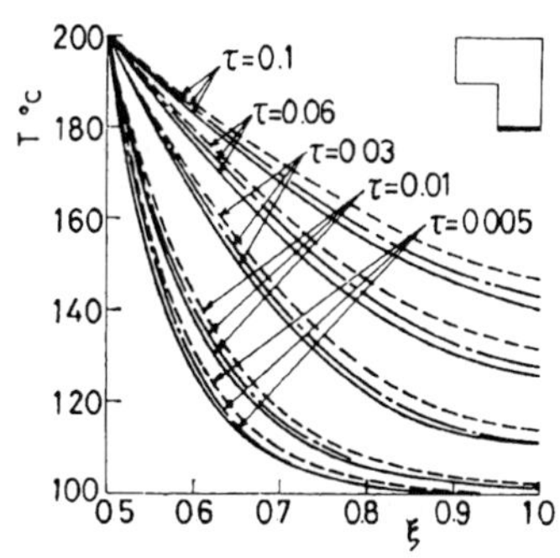

Fig. 3. Temperature distribution on the x-axis as a function of τ for Case (1)

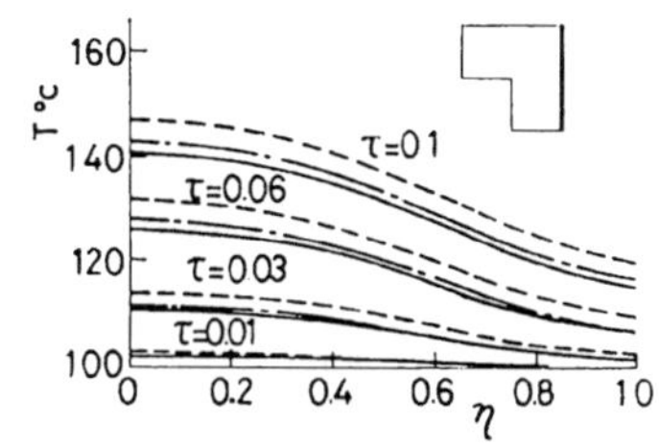

Fig. 4. Temperature distribution on the outer boundary $\xi = 1$ as a function of τ for Case (1)

Figs. 3—7 show the temperature and thermal stress distributions with the dimensionless time τ as a parameter calculated for Case (1). First, Figs. 3 and 4 show the temperature distributions on the x-axis $\eta = 0$ and the outer boundary $\xi = 1$, respectively. As illustrated in Fig. 3, when the temperature dependencies of the thermal properties are taken into account, the temperature in the interior of the plate is lower than that calculated by using the constant thermal properties because the thermal conductivity of low-carbon steel becomes smaller with the increase of the temperature rise; while the temperature gradient is larger because the inner boundary is kept at the constant temperature T_1. If the values of the thermal properties at the initial temperature as the constant thermal properties are used, the temperature distribution on the x-axis at the short period after the beginning of heating is roughly the same whether the variation of the thermal properties with temperature is considered or not. It is noted from Fig. 4 that on the outer boundary $\xi = 1$ far from the inner boundary heated the consideration of the temperature dependencies of the thermal properties yields the lower temperature and the smaller temperature gradient than those of the constant thermal properties. Next, Figs. 5—7 show the distributions of σ_{yy} on the x-axis $\eta = 0$, the inner boundary $\xi = \bar{l}'_x$ and the outer boundary $\xi = 1$, respectively. It is found from these figures that the maximum thermal stress $(\sigma_{yy})_{\max}$ at each time occurs on the inner boundary, and that the absolute value of the maximum

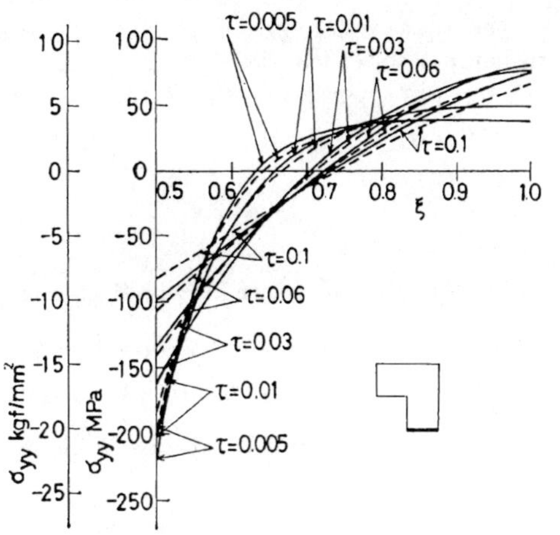

Fig. 5. Effect of the temperature dependencies of the material properties on σ_{yy} on the x-axis for Case (1)

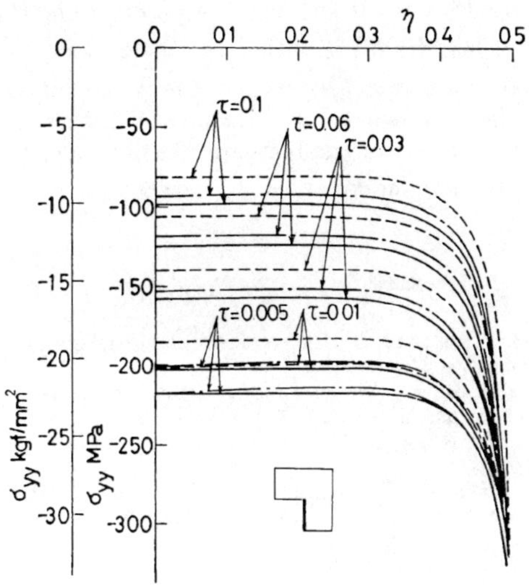

Fig. 6. Effect of the temperature dependencies of the material properties on σ_{yy} on the inner boundary $\xi = \bar{l}_x'$ for Case (1)

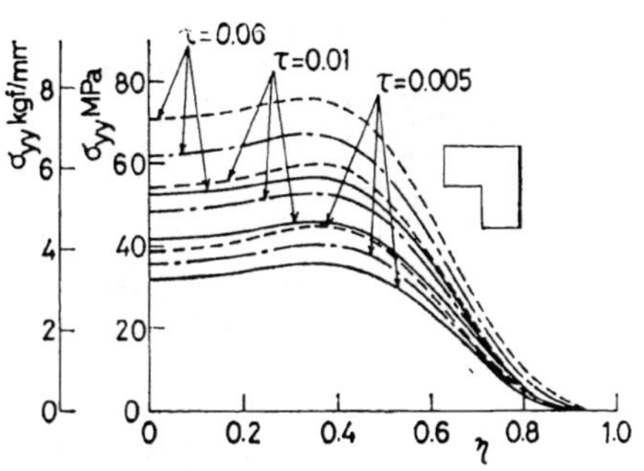

Fig. 7. Effect of the temperature dependencies of the material properties on σ_{yy} on the outer boundary $\xi = 1$ for Case (1)

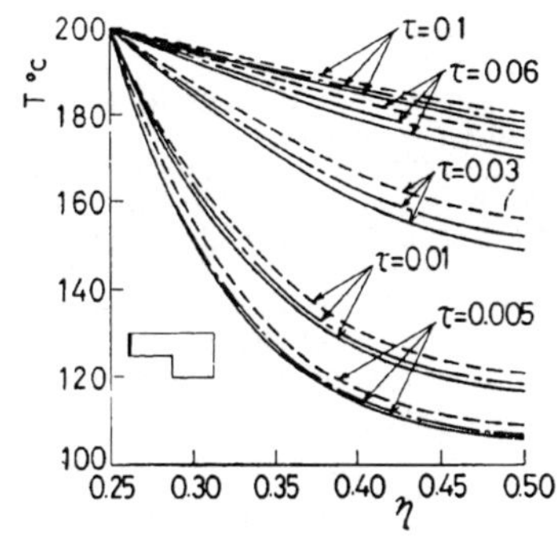

Fig. 8. Temperature distribution on the y-axis as a function of τ for Case (2)

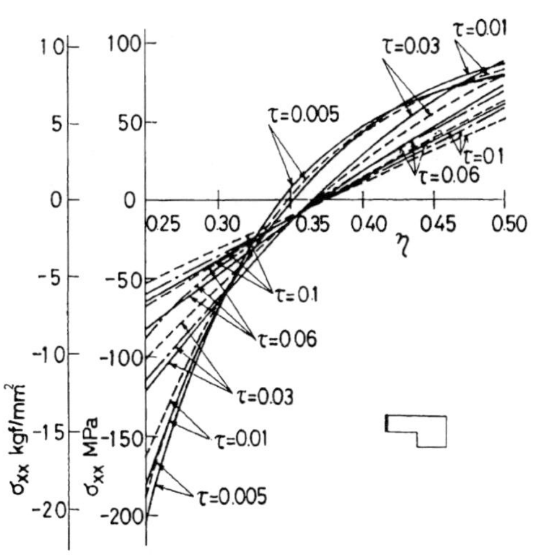

Fig. 9. Effect of the temperature dependencies of the material properties on σ_{xx} on the y-axis for Case (2)

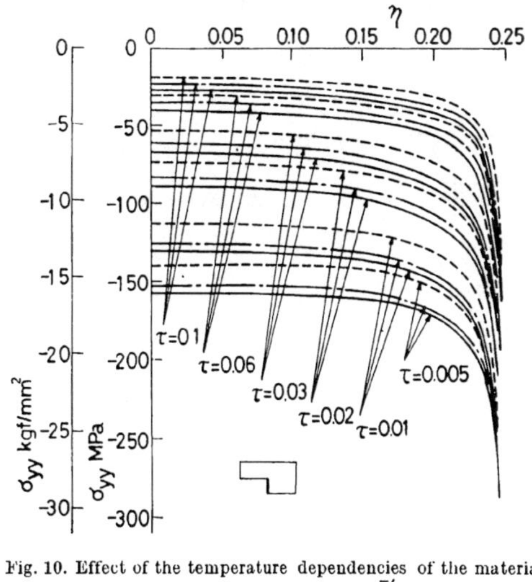

Fig. 10. Effect of the temperature dependencies of the material properties on σ_{yy} on the inner boundary $\xi = \bar{l}_x'$ for Case (2)

stress calculated taking into account the temperature dependencies of the material properties becomes larger than those calculated using the constant material properties. This is due to that the temperature gradient becomes larger by taking into account the temperature dependencies of the thermal properties as shown in Fig. 3, and that the value of the product of Young's modulus and the linear thermal expansion coefficient with the temperature rise becomes larger. It is of practical interest to note from Figs. 6 and 7 that the transient thermal stress σ_{yy} on the inner boundary calculated using the material properties at the initial temperature can express that calculated using the variable material-properties fairly well, but σ_{yy} on the outer boundary obtained using the material properties at the initial temperature does not become a satisfactory approximation to σ_{yy} obtained from the use of the variable material properties.

Finally, Figs. 8—10 show the temperature and thermal-stress distributions on the y-axis and the inner boundary $\xi = \bar{l}_x'$. It is clear from Fig. 9 that the thermal stress σ_{xx} at $\xi = 0$, $\eta = \bar{l}_y'$ on y-axis with the thin thickness becomes smaller than σ_{yy} at $\xi = \bar{l}_x'$, $\eta = 0$ on the x-axis for Case (1) of the square plate because the temperature gradient on the y-axis as shown in Fig. 8 becomes considerably small in comparison with that on the x-axis for Case (1). However, the effects of the temperature dependencies of the material properties on σ_{xx} at this point at $\tau = 0.005$ are similar to that on σ_{yy} at $\xi = \bar{l}_x'$, $\eta = 0$ at the same time of Case (1).

References

1 TAKEUTI, Y.; SEKIYA, T., Thermal Stresses in a Polygonal Cylinder with a Circular Hole under Internal Heat Generation, ZAMM **48** (1968), 237—246.
2 HULBERT, L. E., Solution of Thermal Stress Problems in Tube Sheets by the Boundary Point Least Squares Method, J. Eng. for Industry, Trans. ASME **92** (1970), 339—349.
3 TAKEUTI, Y.; NODA, N., Transient Thermoelastic Problem in a Polygonal Cylinder with a Circular Hole, J. Appl. Mech. **40** (1973), 935—940.
4 SUMI, N.; SEKIYA, T., Thermoelastic Problems for Multiply-Connected Regions with Heat Dissipation at the Both Plane Surfaces, Proc. 20th Japan Nati. Congr. Appl. Mech., (1971), 125—131.
5 TAKEUTI, Y., Thermal Stresses in Heat-Generating Multi-Bore Square or Circular Region, Nucl. Eng. Des. 14 (1970), 201—210.
6 KETTLEBOROUGH, C. F., Non-Axial-Symmetric Thermal Stresses in Circular Discs or Cylinders, Int. Jour. for Num. Math. in Eng. 3 (1971), 53—61.
7 NOWINSKI, J., Transient Thermoelastic Problem for an Infinite Medium with a Spherical Cavity Exhibiting Temperature-Dependent Properties, J. Appl. Mech. 29 (1962), 399—407.
8 NYUKO, H.; TAKEUTI, Y.; NODA, N., Thermal Stresses in a Composite Hollow Circular Cylinder with Temperature Dependency of Material Properties, Trans. Japan Soc. Mech. Engrs. 44 (1978), 1454—1460 (in Japanese).
9 AMADA, S.; YANG, Y. H., Non-Linear Elastic Thermal Stress Analysis with Phase Changes, Nucl. Eng. Des. 48 (1978), 451—460.
10 HATA, T., Thermoelastic Problem for a Griffith Crack in a Plate Whose Shear Modulus is an Exponential Function of the Temperature, ZAMM **61** (1981), 81—87.
11 BOLEY, B. A.; WEINER, J. H., Theory of Thermal Stresses, John Wiley & Sons, New York 1960.
12 MICHELL, J. H., On the Direct Determination of Stress in an Elastic Solid, with Application to the Theory of Plates, Proc. of London Math. Soc. 31 (1899), 100—124.
13 KOBAYASHI, K., Transient Method for Measuring Thermophysical Properties by Stepwise Heating, Jour. Japan Soc. Mech. Engrs. **81** (1978), 452—457 (in Japanese).
14 TAKEUTI, Y., Problems of Thermal Stresses and their Analyses, Jour. Japan Soc. Mech. Engrs. **83** (1980), 279—287 (in Japanese).
15 TAKEUTI, Y., Analyses of Thermal Stresses, Nissin, Tokyo 1971 (in Japanese).

Received 1982 October 21

Address: Dr. Y. SUGANO, Dept. of Mech. Engng., Faculty of Engng, Univ. of Osaka Prefecture, Mozuumemachi, Sakai 591, Osaka, Japan

ZAMM 64, 173 – 191 (1984)

G. Stoyan

Explicit Error Estimates for Difference Schemes Solving the Stationary Constant Coefficient Diffusion-Convection-Reaction Equation

Für die genannte (zeitunabhängige) Gleichung werden Differenzenschemata untersucht und ihre Genauigkeit ausschließlich mit Hilfe verfügbarer Größen abgeschätzt. Die Ergebnisse sind für beliebige Pecletzahlen brauchbar und beinhalten auch gleichmäßige Konvergenz (bezüglich eines kleinen Diffusionskoeffizienten) bis zur vierten Ordnung in h im linearen und von erster Ordnung im schwach nichtlinearen Fall.

Some difference schemes for the cited (stationary) equation are considered and their accuracy is estimated by known quantities only. The results are useful for any Peclét number and imply uniform convergence (with respect to a small diffusion coefficient) up to forth order in h in the linear case and of first order in the mildly nonlinear case.

Рассмотрены разностные схемы решения краевых задач для указанного (стационарного) уравнения и оценена их точность только через известные величины. Результаты пригодны при любом отношении коэффициентов диффузии и конвекции и включают в себя и равномерную сходимость (относительно малого коэффициента диффузии) до четвертого порядка по h в линейном и первого порядка в слабо нелинейном случае.

1. Introduction

We consider the boundary value problem

$$a \frac{\mathrm{d}^2u}{\mathrm{d}x^2} + b \frac{\mathrm{d}u}{\mathrm{d}x} + f(x) = 0\,, \qquad 0 < x < 1\,, \qquad u(0) = l\,, \qquad u(1) = r\,, \tag{1}$$

where f is a sufficiently smooth function, and a and b are constants. We may assume

$$a \geqq 0\,, \qquad b \geqq 0\,, \qquad a + b > 0\,, \tag{2}$$

since $b \geqq 0$ can be brought about by changing x to $1 - x$, if necessary. We shall also consider a mixed boundary value problem for equation (1), and the mildly nonlinear case $f = f(x, u)$. For such problems, due to practical requirements, the construction of difference schemes which are uniformly (with respect to small a) convergent has found much attention recently, see [1]–[7] and further papers therein.

Especially, in [6] it has been shown how to build difference schemes of any prescribed order for the more general equation

$$a \frac{\mathrm{d}^2u}{\mathrm{d}x^2} + b(x) \frac{\mathrm{d}u}{\mathrm{d}x} - d(x)\, u + f(x) = 0$$

under the assumptions $a > 0$, $b(x) \geqq b_0 > 0$, $d(x) \geqq 0$, where b, d and f are piecewise smooth functions.

It seems rather clear and is consistent with numerical experience that schemes of the type [1]–[7] are *general-purpose difference schemes* in the (restricted) sense that they stably define a discrete solution whenever (1) defines (for smooth f) a solution, i.e., if (2) is met.

We shall show here that this is the case, indeed, by proving estimates of the type

$$\|u - y\|_h \leqq \min\left(\frac{1}{a}, \frac{1}{b}\right) M \, \|f^{(n)}\| \, h^n\,,$$

where h is the grid step, u the solution of (1), y the discrete solution, $\|\cdot\|_h$ and $\|\cdot\|$ some norms, and M an explicitely given number. The difference scheme [8] and two schemes of [9] are shown to correspond to $n = 1, 2, 3$, and a forth scheme is constructed for $n = 4$.

A great part of the paper is devoted to the mildly nonlinear case. Here we are close to [10], prove and use the connection between the exact difference scheme [11] and the Ritz method, and also obtain accuracy estimates of the above type, see Theorem 4. Finally, by introducing a norm equivalent to the maximum norm, we show the Nemytzki operator of the problem to be contracting.

2. Notations

We introduce the following notations:

$$x_i := ih\,, \qquad 0 \leqq i \leqq N\,, \qquad h = 1/N\,, \qquad \omega_h := \{x_i,\, i = 1, \ldots, N-1\}\,,$$

$$y_i := y(x_i)\,, \qquad y_{\bar{x},i} := (y_i - y_{i-1})/h\,, \qquad y_{x,i} := (y_{i+1} - y_i)/h\,,$$

$$x_{i+1/2} := x_i + \frac{h}{2}\,, \qquad 0 \leqq i \leqq N-1, \qquad f_{i+1/2} := f(x_{i+1/2})\,,$$

$$\tilde{f}_i := (f_{i+1/2} + f_{i-1/2})/2\,, \qquad f_{\tilde{x},i} := (f_{i+1/2} - f_{i-1/2})/h\,, \qquad f_{\bar{x}x,i} := (f_{i+1} - 2f_i + f_{i-1})/h^2\,.$$

We use the norms

$$||f||_C := \max_{[0,1]} |f(x)| , \qquad ||f|| := \int_0^1 |f(x)|\, dx ,$$

and, for functions of discrete argument,

$$||y||_{C,h} := \max_{\omega_h} |y(x)| , \qquad ||y||_{1,h} := \Big(\sum_{i=1}^{N} (y_{\bar{x},i})^2\, h\Big)^{1/2} .$$

For $n \times n$-matrices $A = (a_{ij})$ we define $||A||_C := \max_{1 \leq i \leq n} \sum_{j=1}^{n} |a_{ij}|$. Further, we shall need the auxiliary functions

$$\varrho(q) := q \coth q , \qquad \alpha(q) := (\varrho(q) - 1)/2q , \qquad \zeta(h) := \coth p - h \coth ph ,$$

where $p := b/2a$, $q := ph$. Some properties of ϱ, α and ζ are collected in the Appendix.

3. The exact difference scheme and some related schemes

For a linear selfadjoint boundary value problem of second order, the exact difference scheme (i.e., the scheme the solution y of which coincides at the grid points with the solution u of the continuous problem) has been defined and investigated in [11], see also [12], pp. 204–207.

The non-selfadjoint problem (1) is first transformed into a selfadjoint one by multiplication with $\exp(bx/a)$, see [7]. After cancelling a common factor $\exp(bx_i/a)$ from the difference equations obtained as in [12], we arrive at the following explicit representation of the exact difference scheme:

$$\Lambda y + \varphi = 0 , \qquad x \in \omega_h; \qquad y_0 = l , \qquad y_N = r . \tag{3); (4}$$

Here Λ is the difference operator

$$(\Lambda y)_i := \frac{1}{h} [(a + bh(\tfrac{1}{2} + \alpha))\, y_{x,i} - (a - bh(\tfrac{1}{2} - \alpha))\, y_{\bar{x},i}] , \tag{5}$$

and $\varphi = \varphi_i$ is the linear functional

$$\varphi_i(f) := \frac{b}{h} \coth q \int_{x_{i-1}}^{x_{i+1}} G_i(x_i, \xi)\, f(\xi)\, d\xi , \tag{6}$$

where G_i denotes Green's function for the first kind boundary value problem of equation (1) on the interval $[x_{i-1}, x_{i+1}]$,

$$G_i(x, \xi) := \frac{2 \exp(p(\xi - x))}{b \sinh 2q} \begin{cases} \sinh p(x - x_{i-1}) \sinh p(x_{i+1} - \xi) , & x \leqq \xi \\ \sinh p(\xi - x_{i-1}) \sinh p(x_{i+1} - x) , & x \geqq \xi \end{cases} . \tag{7}$$

Using the above formulae, we calculate the values of φ_i on the functions x^n, $0 \leqq n \leqq 3$:

$$\varphi_i(1) = 1 , \qquad \varphi_i(x) = x_i + \alpha h , \qquad \varphi_i(x^2) = x_i^2 + 2\alpha h x_i + \frac{1}{3}\left(1 - 3\frac{\alpha}{q}\right) h^2 ,$$
$$\varphi_i(x^3) = x_i^3 + 3\alpha h x_i^2 + \left(1 - 3\frac{\alpha}{q}\right) h^2 x_i + \frac{1}{2}\left(\alpha + \frac{1}{2q}\left(\frac{6\alpha}{q} - 1\right)\right). \tag{8}$$

By inserting suitable approximations $F_i^n(x)$ instead of $f(x)$ into (6), we get several usable difference schemes. We shall follow the lines of [7] and investigate the accuracy of the obtained schemes by comparison with the exact scheme (3)–(7).

Specifically, we are interested in the following F_i^n:

$$n = 1: \quad F_i^1 := f_i ,$$
$$n = 2: \quad F_i^2 := f_i + (x - x_i) \begin{cases} f_{x,i} & b \geqq 0 \\ f_{\bar{x},i} & b \leqq 0 \end{cases} , \quad \tilde{F}_i^2 := \tilde{f}_i + (x - x_i)\, f_{\tilde{x},i} ,$$
$$n = 3: \quad F_i^3 := f_i + (x - x_i)\, \tfrac{1}{2} (f_{\bar{x},i} + f_{x,i}) + \tfrac{1}{2} (x - x_i)^2 f_{\bar{x}x,i} .$$

With the help of the identities (8) we get explicit expressions for $\varphi_i(F_i^n) =: \varphi_i^n$, $\varphi_i(\tilde{F}_i^2) =: \tilde{\varphi}_i^2$:

$$\varphi_i^1 = f_i , \qquad \varphi_i^2 = \max(0, \alpha)\, f_{i+1} + (1 - |\alpha|)\, f_i + \max(0, -\alpha)\, f_{i-1} ,$$
$$\tilde{\varphi}_i^2 = \left(\frac{1}{2} + \alpha\right) f_{i+1/2} + \left(\frac{1}{2} - \alpha\right) f_{i-1/2} , \qquad \varphi_i^3 = \varphi_i^2 + \frac{h^2}{2}\left(\frac{1}{3} - \frac{\alpha}{q}(1 + |q|)\right) f_{\bar{x}x,i} . \tag{9}$$

For the difference schemes (3) with $\varphi_i = \varphi_i^n$ see [8] ($n = 1$), [9] ($n = 2$, $n = 3$; comp. p. 363 and Remark 1 at p. 364). The scheme with $\varphi_i = \tilde{\varphi}_i^2$ is a special case of [13].

In the above formulae, we have tacitly assumed $a > 0$, $b > 0$, but the limiting cases $a > 0$, $b = 0$, and $= 0$, $b > 0$ are included if defined as result of letting $b \to 0$ or $a \to 0$ (i.e., $q \to 0$ or $q \to \infty$), see also the Appendix.

In what follows, we suppose $a > 0$, $b > 0$, and the statements below will be seen to be senseful and true in the limiting cases, too.

4. The Green's function

To obtain accuracy estimates for the schemes (3)—(5), (9), we shall consider the discrete Green's function $G^h(x_i, \xi_j) = G^h_{ij}$, that is, the function which satisfies (as a function of x_i) the equation (3) with $\varphi_i = \frac{1}{h} \delta_{ij}$, δ_{ij} the Kronecker symbol, and hence lends itself to the representation

$$y_i = \sum_{j=1}^{N-1} G^h_{ij} \varphi_i(j)\, h \tag{10}$$

for the solution y of (3), in case $l = r = 0$.

Following [7] we find

$$G^h_{ij} = \frac{1}{b\varkappa^j(\varkappa^{-N} - 1)} \begin{cases} (1 - \varkappa^i)(\varkappa^{j-N} - 1), & i \leqq j \\ (1 - \varkappa^j)(\varkappa^{i-N} - 1), & i \geqq j, \end{cases} \tag{11}$$

where $\varkappa := (a - bh(\frac{1}{2} - \alpha))/(a + bh(\frac{1}{2} + \alpha)) = e^{-2q} \in (0, 1)$. An equivalent form is

$$G^h(x_i, \xi_j) = \frac{2 \exp(p(\xi_j - x_i))}{b \sinh p} \begin{cases} \sinh p x_i \sinh p(1 - \xi_j), & x_i \leqq \xi_j \\ \sinh p\xi_j \sinh p(1 - x_i), & x_i \geqq \xi_j, \end{cases} \tag{12}$$

which shows that $G^h(x, \xi)$ equals $G(x, \xi)$, the Green's function of (1), at the grid points. More is true:

Due to $\varphi_i(1) = 1$, the problem

$$aY'' + bY' + 1 = 0, \qquad 0 < x < 1, \qquad Y(0) = Y(1) = 0, \tag{13}$$

is solved exactly by the scheme $\Lambda Y + 1 = 0$, $x \in \omega_h$, $Y_0 = Y_N = 0$. This and (10) imply

$$\int_0^1 G(x_i, \xi)\, d\xi = Y(x_i) = \sum_{j=1}^{N-1} G^h_{ij} h = \sum_{j=1}^{N-1} G(x_i, \xi_j)\, h\,.$$

As can be easily verified, the solution to (13) is

$$Y(x) = \frac{1}{b}\left[1 - x + \frac{1}{2\sinh p}(e^{\,p} - e^{p(1-2x)})\right]. \tag{14}$$

Lemma 1: *The following inequalities are true:*

$$0 \leqq G^h_{ij} \leqq \min\left(\frac{1}{4a}, \frac{1}{b}\right), \qquad \sum_{j=1}^{N-1} G^h_{ij} h \leqq \min\left(\frac{1}{8a}, \frac{1}{b}\right),$$

$$\sum_{j=1}^{N-1} |G^h_{\bar{x},ij}|\, h \leqq \begin{cases} \min\left(\frac{1}{a}, \frac{1}{b}\right), & \frac{N}{2} \leqq i \leqq N \\ \min\left(\frac{1}{a}, \frac{1}{bh}\right), & 1 \leqq i < \frac{N}{2} \end{cases}. \qquad \sum_{i=1}^{N} \sum_{j=1}^{N-1} (G^h_{\bar{x},ij})^2 h^2 \leqq \min\left(\frac{1}{6a^2}, \frac{2}{hb^2}\right).$$

Proof: Since the lemma is based on the explicit representations (11), (12), we shall not prove all inequalities. To obtain the second inequality, we first determine the maximum of $Y(x)$:

$$\frac{1}{2b}\left[\frac{1}{p} \ln \frac{\sinh p}{p} + 2\alpha(p)\right] = \max_{[0,1]} Y(x) \geqq \sum_{j=1}^{N-1} G^h_{ij} h\,.$$

We take $2\alpha(p) \leqq \min\left(\frac{p}{3}, 1\right)$ from the Appendix and estimate $P := \frac{1}{p} \ln \frac{\sinh p}{p}$ as follows: Firstly, $P \leqq 1$:

$$\sinh p = \tfrac{1}{2}(e^p - e^{-p}) \leqq \tfrac{1}{2}(1 + p\, e^p - (1 - p)) \leqq p\, e^p\,.$$

Secondly, $P \leqq \frac{p}{6}$, which can be verified easily by comparing the power series of $\sinh p$ and $p \exp(p^2/6)$. Hence $P \leqq \min\left(1, \frac{p}{6}\right)$, and the second inequality follows. By the way,

$$\max_{[0,1]} Y(x) = \frac{1}{8a} \quad \text{for} \quad a > 0, \quad b = 0, \quad \text{and} \quad \max_{[0,1]} Y(x) = \frac{1}{b} \quad \text{for} \quad b > 0, \quad a = 0\,.$$

To prove the third estimate in case $i \geqq N/2$, we observe that

$$G^h_{\bar{x},ij} > 0, \quad x_i \leqq \xi_j, \quad G^h_{\bar{x},ij} < 0, \quad x_i \geqq \xi_j + h\,.$$

Therefore,

$$\sum_{j=1}^{N-1} |G^h_{\bar{x},ij}|\, h = \sum_{j=1}^{i} G^h_{ij} + \sum_{i+1}^{N-1} G^h_{i+1j} - \sum_{j=1}^{i} G^h_{i+1j} - \sum_{i+1}^{N-1} G^h_{ij}\,.$$

Performing the summation, we get

$$\sum_{j=1}^{N-1} |G^h_{\bar{x},ij}|\, h = \frac{1}{b}\left[\frac{e^{2p}+1}{e^{2p}-1} + \frac{e^{2(p-qi)}}{e^{2p}-1}\left((N-2i)\,(1-e^{-2q})-2\right)\right] = \frac{1}{b}\left[\tanh\frac{p}{2} + S\right],$$

where

$$S := \frac{1}{\sinh p}\left[-s\frac{e^{-s}}{\varrho+q} + 1 - e^{-s}\right], \quad s := p(2x-1)\,, \quad \varrho(q) := q\coth q.$$

Now, for $\frac{1}{2} \leqq x_i \leqq 1$, we have, firstly, $-s\frac{e^{-s}}{\varrho+q} \leqq 0$ and $1-e^{-s} \leqq 1-e^{-p}$. Hence,

$$\sum_{j=1}^{N-1} |G^h_{\bar{x}i,j}|\, h \leqq \frac{1}{b}\left[\tanh\frac{p}{2} + \frac{1-e^{-p}}{\sinh p}\right] = \frac{1}{b}.$$

Secondly, with the help of the elementary inequality $-e^{-s} \leqq -\left(1-\frac{s^2}{2}\right)/(1+s)$, $s \geqq 0$, we may estimate S as follows:

$$S \leqq \frac{1}{(1+s)\sinh p}\left[\frac{-s}{\varrho+q}\left(1-\frac{s^2}{2}\right) + s + \frac{s^2}{2}\right] = \frac{s\left[-1+\frac{s^2}{2}+\left(1+\frac{s}{2}\right)(1+hp(1+2\alpha))\right]}{(1+s)\sinh p\,(1+hp(1+2\alpha))} =$$

$$= \frac{s\left[\frac{s^2}{2}+\frac{s}{2}+\left(1+\frac{s}{2}\right)hp\,(1+2\alpha)\right]}{(1+s)\sinh p\;(1+hp(1+2\alpha))}\,,$$

because of $\varrho = 1+2q\alpha = 1+2hp\alpha$. Using $s \leqq p$, $\sinh p \geqq p$, $h \leqq \frac{1}{2}$ and $0 \leqq \alpha \leqq \frac{1}{2}$ (see the Appendix), we obtain from here $S \leqq \frac{3}{2}p$, and $\tanh\frac{p}{2} \leqq \frac{p}{2}$ leads then to the required estimate.

As for the last inequality, we find

$$\sum_{i=1}^{N}\sum_{j=1}^{N-1} (G^h_{\bar{x},ij})^2\, h^2 = \frac{2}{hb^2}\zeta(h)\tanh q$$

by straightforward calculation. Exploiting

$$\zeta(h) \leqq 2\alpha(p) \leqq \min\left(\frac{p}{3}, 1\right)$$

from the Appendix and using also $\tanh q \leqq \min(q, 1)$, we get the assertion of the lemma.

Next we turn to the functional $\varphi_i(f)$ defined by (6) and (7).

Lemma 2: *We have*

$$|\varphi_i(f)| \leqq \max_{[x_{i-1},\,x_{i+1}]} |f(x)| \quad \text{and} \quad |\varphi_i(f)| \leqq \frac{1}{h}\int_{x_{i-1}}^{x_{i+1}} |f(\xi)|\,d\xi\,.$$

Proof: The first inequality follows since φ_i is an isotone linear functional on $C[x_{i-1}, x_{i+1}]$ satisfying $\varphi_i(1) = 1$. Further, we remark that

$$0 \leqq G_i(x,\xi) \leqq G_i(\xi,\xi) \leqq \frac{1}{b}\tanh q\,,$$

which implies the second inequality.

5. Stability

We prove now an estimate for the scheme (3)—(5) showing its maximum norm stability for arbitrary right-hand sides φ_i with $\max\limits_{1\leqq i\leqq N-1} |\varphi_i| \leqq$ const.

Theorem 1: *For the solution y of (3)—(5) the following estimate holds:*

$$\|y\|_{C,h} \leqq \max(|l|, |r|) + \min\left(\frac{1}{8a}, \frac{1}{b}\right)\max_{1\leqq i\leqq N-1} |\varphi_i|\,.$$

Proof: We have $a \pm bh(\frac{1}{2} \pm \alpha) = a(\varrho \pm q) \geqq 0$ and $a + bh(\frac{1}{2}+\alpha) + a - bh(\frac{1}{2}-\alpha) = 2(a + bh\alpha)$, where $b\alpha \geqq 0$ (with equality only in case $b = 0$, see the Appendix). Now the maximum principle (cf. [12], p. 245) gives immediately the above estimate for $\varphi_i = 0$, $1 \leqq i \leqq N-1$.

For the inhomogeneous equation (3) with homogeneous boundary conditions we use the majorizing function $mY(x)$ — see [12], p. 247, for this method — with $m > 0$ and Y defined by (13)—(14), i.e.

$$Y_i = \sum_{j=1}^{N-1} G_{ij}^h h .$$

We have then

$$-(\Lambda m Y)_i = m \geqq |\varphi_i| ,$$

if $m = \max_{1 \leqq j \leqq N-1} |\varphi_j|$ is selected. Hence

$$||y||_{C,h} \leqq ||mY||_{C,h} \leqq \min\left(\frac{1}{8a}, \frac{1}{b}\right) m , \quad \text{by Lemma 1} .$$

Remark: Let be A_Λ the $(N-1) \times (N-1)$-matrix corresponding to the difference operator $-\Lambda$ under homogeneous boundary conditions. Then A_Λ^{-1} has the elements $G_{ij}^h h$, see (10). Therefore,

$$||A_\Lambda^{-1}||_C = \max_{1 \leqq i \leqq N-1} \sum_{j=1}^{N-1} G_{ij}^h h = ||Y||_{C,h} ,$$

and, since $||A_\Lambda^{-1}||_C \geqq \lambda_{\max}^h(A_\Lambda^{-1}) = 1/\lambda_1^h$ for the greatest eigenvalue $\lambda_{\max}^h$ of A_Λ^{-1} and the first eigenvalue λ_1^h of A_Λ, it follows that

$$1/||Y||_C \leqq 1/||Y||_{C,h} \leqq \lambda_1^h .$$

Explicit expressions for the eigenvalues and eigenfunctions of $-\Lambda$, i.e. of A_Λ, have been given in [14]. Especially,

$$\lambda_1^h = \frac{2a}{h^2}\varrho(q)\left(1 - \frac{\cos \pi h}{\cosh q}\right), \qquad \text{and} \quad \lambda_1^h < \lambda_1 = a\pi^2 + \frac{b^2}{4a}$$

is easily proved for the first eigenvalue λ_1 of the continuous problem.

6. Accuracy estimates

Let y^n and $\tilde{y}^2$ denote the solution of (3)—(5) with right-hand side $\varphi_i = \varphi_i^n$, $n = 1, 2, 3$, and $\varphi_i = \tilde{\varphi}_i^2$, respectively. The Lemmas 1 and 2 will be used now to estimate the accuracy of these approximations to the solution $u(x)$ of (1).

Theorem 2: *For the solutions of* (3)—(5), (9), *the following inequalities are true:*

$$||u - y^n||_{C,h} \leqq \min\left(\frac{1}{4a}, \frac{1}{b}\right) 2h^n\, ||f^{(n)}|| , \qquad n = 1, 2 ,$$

$$||u - \tilde{y}^2||_{C,h} \leqq \min\left(\frac{1}{4a}, \frac{1}{b}\right) h^2 ||f''|| , \qquad ||u - y^3||_{C,h} \leqq \min\left(\frac{1}{8a}, \frac{1}{b}\right)\frac{h^3}{2} ||f'''||_C .$$

Proof: Let be $z^n = y - y^n$, where y is the solution of the exact scheme (3)—(7), $y_i = u(x)_i$. Then $z_0 = z_N = 0$ and

$$z_i^n = \sum_{j=1}^{N-1} G_{ij}^h \varphi_j (f - F_j^n)\, h .$$

Hence

$$|z_i^n| \leqq \min\left(\frac{1}{4a}, \frac{1}{b}\right) \sum_{j=1}^{N-1} \max_{[x_{j-1}, x_{j+1}]} |f - F_j^n|\, h ,$$

by the Lemmas 1 and 2.

We assume $f \in C^{n-1}[0, 1]$, $f^{(n)} \in L^1[0, 1]$ and obtain for $n = 1, 2$

$$f(x) - F_j^n(x) = \int_{x_{j-1}}^{x_{j+1}} g_j^n(x, x_j, \xi)\, f^{(n)}\, d\xi , \qquad f(x) - \tilde{F}_j^2(x) = \int_{x_{j-1}}^{x_{j+1}} \tilde{g}_j^2(x, x_j, \xi)\, f''\, d\xi$$

from the Taylor development, where g_j^1 is a piecewise continuous function satisfying $|g_j^1| \leqq 1$, whereas g_j^2 and $\tilde{g}_j^2$ are continuous functions with $|g_j^2| \leqq h$, $|\tilde{g}_j^2| \leqq \frac{h}{2}$ for $x \in [x_{j-1}, x_{j+1}]$. Therefore,

$$\max_{[x_{j-1}, x_{j+1}]} |f - F_j^n| \leqq h^{n-1} \int_{x_{j-1}}^{x_{j+1}} |f^{(n)}|\, d\xi , \qquad \max_{[x_{j-1}, x_{j+1}]} |f - \tilde{F}_j^2| \leqq \frac{h}{2} \int_{x_{j-1}}^{x_{j+1}} |f''|\, d\xi .$$

Now the first three inequalities follow from

$$\sum_{j=1}^{N-1} \int_{x_{j-1}}^{x_{j+1}} |f^{(n)}|\, d\xi \leqq 2||f^{(n)}|| .$$

For $n = 3$ we use the second inequality of Lemma 1, the first of Lemma 2, and

$$\max_{[x_{j-1}, x_{j+1}]} |f - F_j^3| \leqq \frac{h^3}{2} ||f'''||_C .$$

Remark: The third and forth inequality of Lemma 1 lead to similar explicit error estimates for $y_{\bar{x}}^n$ and $\tilde{y}_{\bar{x}}^2$:

$$||y - y^n||_{1,h} \leqq \min\left(\frac{1}{6a}\sqrt{6}\,, \frac{1}{b}\sqrt{\frac{2}{h}}\right) M_n h^n ||f^{(n)}||_C\,,$$

$$|y_{\bar{x},i} - y_{\bar{x},i}^n| \leqq M_n h^n ||f^{(n)}||_C \begin{cases} \min\left(\dfrac{1}{a}, \dfrac{1}{b}\right) & \dfrac{N}{2} \leqq i \leqq N\,, \\ \min\left(\dfrac{1}{a}, \dfrac{1}{bh}\right) & 1 \leqq i < \dfrac{N}{2}\,, \end{cases}$$

where $M_1 = M_2 = 2$, $\widetilde{M}_2 = 1$, $M_3 = \frac{1}{2}$. These and the above results (as well as the estimate of Theorem 4 below concerning $f = f(x, u)$) constitute a development of a part of the results obtained in [15] and [16], where always $a = 1$ was considered (but f was assumed to depend on u and u'). We observe that $y_{\bar{x},i} = u'(x_{i-1/2} + \vartheta h)$, $|\vartheta| < \frac{1}{2}$.

7. A fourth order general-purpose scheme

Simple examples show that the order of accuracy of the considered schemes (for $n = 1, 3$) is restricted to the order indicated by Theorem 2 for the case $a = 0$, $b > 0$. But if $a > 0$, $b \to 0$, then the accuracy of these schemes improves by one order for sufficiently smooth f whereas the accuracy of the schemes with $n = 2$ remains $O(h^2)$. For instance, for the scheme with $n = 1$ we have

$$\varphi_i(f - F_i^1) = f_i + \alpha h f_i' + R_2\,, \qquad |R_2| \leqq \frac{h^2}{2} ||f''||_C\,.$$

Using the second inequality of Lemma 1, and $0 \leqq \alpha \leqq \min\left(\frac{q}{6}, \frac{1}{2}\right)$ from the Appendix, we get

$$||u - y^1||_C \leqq \min\left(\frac{1}{8a}, \frac{1}{b}\right)\left[\min\left(\frac{bh}{12a}, \frac{1}{2}\right) h ||f'||_C + \frac{h^2}{2} ||f''||_C\right].$$

Similarly, in the case $n = 3$ the scheme (3), (5), (9) is of fourth order for $a > 0$, $b = 0$ (comp. [9], p. 363): Assuming $f \in C^4[0, 1]$, we calculate

$$f - F_i^3 = \frac{1}{6}(x - x_i)\,[(x - x_i)^2 + h^2]\, f_i''' + R_4\,, \qquad |R_4| \leqq \frac{h^4}{8} ||f^{(4)}||_C\,,$$

and obtain, using (8),

$$\varphi_i\left(\frac{1}{6}(x - x_i)\,[(x - x_i)^2 + h^2]\right) = \frac{h^3}{4}\left(\alpha + \frac{1}{6q}\left(\frac{6\alpha}{q} - 1\right)\right). \tag{15}$$

Here the right-hand side is $O(h^3 q) = O(h^4 b/a)$ for small q, and converges to $\frac{1}{8}h^3 \operatorname{sign} b$ for $q \to \infty$ (see the Appendix).

Exploiting (15), we get a fourth order scheme and may add to (9) the case $n = 4$,

$$\begin{aligned} \varphi_i^4 &:= \varphi_i^3 + \frac{h^3}{4}\left(\alpha + \frac{1}{6q}\left(\frac{6\alpha}{q} - 1\right)\right) f_{3,i}\,, \\ f_{3,i} &:= \begin{cases} f_{\bar{x}xx,i}, & b \geqq 0\,, \quad 1 \leqq i \leqq N-2 \ \text{ and } \ b < 0\,, \quad i = 1 \\ f_{\bar{x}x\bar{x},i}, & b < 0\,, \quad 2 \leqq i \leqq N-1 \ \text{ and } \ b \geqq 0\,, \quad i = N-1\,. \end{cases} \end{aligned} \tag{9}$$

The solution y^4 of (3)—(5), (9), $n = 4$, satisfies the estimate

$$||u - y^4||_{C,h} \leqq \min\left(\frac{1}{8a}, \frac{1}{b}\right) \frac{h^4}{8} ||f^{(4)}||_C\,.$$

8. A mixed boundary value problem

We consider now another case of practical interest: equation (1) along with the boundary conditions

$$-au'(0) = l\,, \qquad u(1) = r\,. \tag{16}$$

The exact difference scheme for (1), (16) is defined by (3), (5)—(7) and the following additional relations for $i = 0$ and $i = N$:

$$(\Lambda y)_0 + \varphi_0 + \frac{l}{h} = 0\,, \qquad y_N = r\,, \tag{17}$$

where

$$(\Lambda y)_0 := \frac{1}{h}\left(a + bh\left(\frac{1}{2}+\alpha\right)\right) y_{x,0} \qquad \text{(comp. [17])},$$

$$\varphi_0 = \varphi_0(f) := \frac{b}{2h}(\coth q + 1)\int_0^h G(0,\xi) f(\xi)\,d\xi,$$

$$G(x,\xi) := \frac{2}{b} e^{p(\xi-h)} \begin{cases} \sinh p(h-\xi), & 0 \leqq x \leqq \xi \leqq h \\ e^{p(\xi-h)} \sinh p(h-x), & 0 \leqq \xi \leqq x \leqq h. \end{cases}$$

For the Green's function of (3), (17) we find the representation

$$G_{ij}^h = \frac{2}{\mathrm{b}(\coth q + 1)} \begin{cases} \sum\limits_{j}^{N-1} \varkappa^{k-j}, & i \leqq j \\ \sum\limits_{i}^{N-1} \varkappa^{k-j}, & i \geqq j, \end{cases} \tag{18}$$

where $\varkappa = (a - bh(\frac{1}{2}-\alpha))/(a + bh(\frac{1}{2}+\alpha)) = e^{-2q}$ as above.

Lemma 3: *The Green's function* (18) *satisfies the following inequalities*:

$$0 \leqq G_{ij}^h \leqq \min\left(\frac{1}{a}, \frac{1}{b}\right), \qquad \sum_{j=0}^{N-1} |G_{\bar{x},ij}^h|\, h \leqq \min\left(\frac{1}{a} x_i, \frac{1}{b}\right).$$

Proof: From (18) we have

$$0 \leqq G_{ij}^h \leqq G_{00}^h = \frac{2}{b(\coth p + 1)},$$

and $\coth p \leqq \max\left(\frac{1}{p}, 1\right)$ leads then to $G_{ij}^h \leqq \min\left(\frac{1}{b}, \frac{1}{a+b/2}\right)$. Further,

$$\sum_{j=0}^{N-1} |G_{\bar{x},ij}^h|\, h = \frac{2}{bh(\coth q + 1)} \frac{1-\varkappa^i}{1-\varkappa} = \frac{1}{b}(1 - e^{-2qi}) \leqq \min\left(\frac{2qi}{b}, \frac{1}{b}\right).$$

We compute the values of the functional $\varphi_0(f)$ on x^n, $n = 0, 1, 2$:

$$\varphi_0(1) = \frac{1}{2}+\alpha, \qquad \varphi_0(x) = \frac{h}{2}\left(\frac{1}{2}+\alpha-\frac{\alpha}{q}\right), \qquad \varphi_0(x^2) = \frac{h^2}{6}\left[1 + \frac{\alpha}{q}(2q-1) + \frac{1}{2q}\left(\frac{6\alpha}{q}-1\right)\right]. \tag{19}$$

Analogously to the first inequality of Lemma 2, we have now

$$|\varphi_0(f)| \leqq \max_{[0,h]} |f(x)|\,(\tfrac{1}{2}+\alpha) \leqq \max_{[0,h]} |f(x)|, \tag{20}$$

and, paralleling (9), we define

$$\varphi_0^1 := \varphi_0(0) = 0, \qquad \varphi_0^2 := \varphi_0(f_0) = f_0(\tfrac{1}{2}+\alpha),$$

$$\tilde{\varphi}_0^2 := \varphi_0(f_{1/2}) = f_{1/2}\left(\frac{1}{2}+\alpha\right), \qquad \varphi_0^3 := \varphi_0(f_0 + x f_{x,0}) = \frac{1}{2}(f_0+f_1)\left(\frac{1}{2}+\alpha\right) - (f_1 - f_0)\frac{\alpha}{2q}, \tag{21}$$

$$\varphi_0^4 := \varphi_0(f_0 + x f_{x,0} + \tfrac{1}{2}x(x-h) f_{\bar{x}x,1})$$

$$= f_0\left(\frac{5}{12}\left(\frac{1}{2}+\alpha\right) + \frac{2\alpha}{3q} + \frac{1}{24q}\left(\frac{6\alpha}{q}-1\right)\right) + f_1\left(\frac{2}{3}\left(\frac{1}{2}+\alpha\right) - \frac{5\alpha}{6q} - \frac{1}{12q}\left(\frac{6\alpha}{q}-1\right)\right) +$$

$$+ f_2\left(-\frac{1}{12}\left(\frac{1}{2}+\alpha\right) + \frac{\alpha}{6q} + \frac{1}{24q}\left(\frac{6\alpha}{q}-1\right)\right).$$

From the representation

$$y_i = r + \sum_{j=1}^{N-1} G_{ij}^h \varphi_j h + G_{i0}\left(\varphi_0 + \frac{l}{h}\right) h$$

of the solution y of (3), (5), (17) we obtain, using Lemma 3, the stability estimate

$$\|y\|_{C,h} \leqq |r| + \min\left(\frac{1}{a}, \frac{1}{b}\right)\left(\sum_{j=0}^{N-1} |\varphi_j|\, h + |l|\right).$$

Concerning accuracy, we formulate the following theorem.

Theorem 3: *The solutions y^n and $\tilde{y}^2$ of the schemes (3), (5), (17), (21) satisfy the following estimates:*

$$\|u - y^n\|_{C,h} \leqq \min\left(\frac{1}{a}, \frac{1}{b}\right) h^n(\|f^{(n-1)}\|_C + 2\|f^{(n)}\|) , \qquad n = 1, 2 ,$$

$$\|u - \tilde{y}^2\|_{C,h} \leqq \min\left(\frac{1}{a}, \frac{1}{b}\right) h^2\left(\frac{1}{2}\|f'\|_C + \|f''\|\right) ,$$

$$\|u - y^3\|_{C,h} \leqq \min\left(\frac{1}{a}, \frac{1}{b}\right) h^3\left(\frac{3}{8}\|f''\|_C + \frac{1}{2}\|f'''\|_C\right) ,$$

$$\|u - y^4\|_{C,h} \leqq \min\left(\frac{1}{a}, \frac{1}{b}\right) h^4\left(\frac{1}{4}\|f'''\|_C + \frac{1}{8}\|f^{(4)}\|_C\right) .$$

Proof: Let be $z^n = y - y^n$ where y is the solution of the exact difference scheme (3)—(7), (17); $y_i = u(x_i)$. Then

$$z_i^n = \sum_{j=0}^{N-1} G_{ij}^h \varphi_j(f - F_j^n)\, h \quad \text{and} \quad |z_i^n| \leqq \max_{i,j} G_{ij}^h(|f - F_0^n|\, h + \sum_{j=1}^{N-1} |f - F_j^n|\, h) ,$$

by Lemma 2 and (20). Now Lemma 3 is applied, the estimate of the sum can be taken over from Theorem 2, and $|f - F_0^n|\, h$ is estimated separately (where F_0^n is defined implicitly by (21), comp. (9)).

Remarks:

1. As usual, for the second kind boundary condition we have chosen approximations the order of which is by one lower than for the inner equations. If we want to exploit the second part of Lemma 3 for estimation of $\|y_{\bar{x}}\|_{C,h}$, then it becomes important to have approximations of equal order everywhere. We formulate one such result: *The solution $\bar{y}^2$ of the difference scheme* (3), (5), (9), $n = 2$, (17), (21), $n = 3$, *satisfies the estimate*

$$\|y_{\bar{x}} - \bar{y}_{\bar{x}}^2\|_{C,h} \leqq \min\left(\frac{1}{a}, \frac{1}{b}\right) 2h^2\, \|f''\|_C .$$

2. From between the considered schemes, the scheme (3), (5), (9), (17), $n = 2$, (21), $\varphi_i = \tilde{\varphi}_i^2$, as well as the scheme of the above Remark 1 seem to be the most satisfactory to the author: These difference schemes are of second order, the weights multiplying the f-values in $\tilde{\varphi}_i^2$, φ_i^2 and φ_0^3 are nonnegative, moreover, these right-hand sides do not use such information on f which, for $a \to 0$, lies outside of the physical domain of dependence, finally, in the limit case $a = 0$ the same formulae arise at the inner and at the boundary point x_0 (if $l = 0$ in (16)). Thus, there will be no numerical boundary layer at all then, and (17) with $l = 0$ can be used to model an outflow boundary.

9. The Ritz method and the exact difference scheme

Befor turning to the mildly nonlinear case, we investigate the relation between the exact difference scheme and a special Ritz method as applied to the selfadjoint boundary value problem

$$\mathcal{L}u + f(x) = 0 , \quad 0 < x < 1 , \quad u(0) = u(1) = 0 , \quad \mathcal{L}u := (a(x)\, u')' + d(x)\, u , \tag{22}$$

where $a(x)$, $d(x)$ and $f(x)$ are piecewise continuous functions satisfying

$$a(x) > 0 , \quad d(x) \leqq \lambda_1 - \varepsilon , \quad 0 \leqq x \leqq 1 , \tag{23}$$

where ε is a small positive number and λ_1 (> 0) the first eigenvalue of $\mathcal{L}_0$, $\mathcal{L}_0 u := -(a(x)\, u')'$,

$$\lambda_1 := \min_{\substack{v \in S \\ v \neq 0}} \left\{\int_0^1 a(x)\, v'^2\, dx \Big/ \int_0^1 v^2\, dx\right\} . \tag{24}$$

Here $S = \mathring{W}_1^2[0, 1]$ is the linear space of absolutely continuous real functions $v(x)$ satisfying $v'(x) \in L^2[0, 1]$ and $v(0) = v(1) = 0$. The solution $u(x)$ of (22) is characterized as (cf. [18]) the unique minimizer on S of the functional

$$F(v) := \int_0^1 [a(x)\, v'^2 - d(x)\, v^2 - 2\int_0^{v(x)} f(\xi)\, d\xi]\, dx .$$

An approximate solution is defined by minimizing F over the subspace S_{N-1} of S consisting of all functions w of the form

$$w(r) = [w(r)]\,(x) := \sum_1^{N-1} r_i w_i(x) ,$$

with real numbers r_i and basis functions $w_i(x)$, $1 \leqq i \leqq N - 1$, defined uniquely by (comp. [19])

$$w_i(x_j) = \delta_{ij} , \quad \mathcal{L}w_i(x) = 0 , \quad 0 < x < 1 , \quad x \neq x_j , \quad j = 0, \ldots, N . \tag{25}$$

Here the knots x_j are (only in this point) nonequidistantly spread over the interval [0, 1] so as to cover the points 0 and 1 and all points of discontinuity of $a(x)$, $d(x)$ and $f(x)$. In the terminology of [20], $w(r)$ is an "*$\mathcal{L}$-spline*".

Now let be $\hat{w}(x)$ the unique minimizer (cf. [18]) of $F(v)$ over S_{N-1}, then

$$\hat{w}(x) = w(y) = \sum_{1}^{N-1} y_i w_i(x) \,, \tag{26}$$

where the $(N-1)$-dimensional vector $y = (y_1, \dots, y_{N-1})^{\mathsf{T}}$ is determined by the system of linear equations

$$Ay = \varphi \,, \tag{27}$$

with

$$A = (a_{ij}) \,, \; a_{ij} := \int_0^1 [a(x)\, w_i' w_j' - d(x)\, w_i w_j]\, \mathrm{d}x \,, \quad \varphi = (\varphi_1, \dots, \varphi_{N-1})^{\mathsf{T}} \,, \; \varphi_i := \int_0^1 w_i(x)\, f(x)\, \mathrm{d}x \,.$$

The matrix A is, obviously, symmetric. Moreover, A is also positive definite: For $(N-1)$-dimensional vectors z we have

$$\sum_{i=1}^{N-1} \sum_{j=1}^{N-1} a_{ij} z_i z_j =: (Az, z) = \int_0^1 [a(x) \, (\sum_1^{N-1} z_i w_i')^2 - d(x) \, (\sum_1^{N-1} z_i w_i)^2]\, \mathrm{d}x \,,$$

and, using (23) and (24), we get

$$(Az, z) \geqq \varepsilon \int_0^1 (\sum_1^{N-1} z_i w_i)^2 \, \mathrm{d}x \,.$$

Now put $v(x) = (\sum_1^{N-1} z_i w_i)^2$ and assume $z_j \neq 0$. Then $v(x_j) = z_j^2 > 0$ and, by the continuity of v, there is a neighbourhood of x_j in which $v(x) > 0$. Hence $(Az, z) \geqq 0$ with equality only if $z = (0, \dots, 0)$.

To obtain the basis functions w_i, we must solve boundary value problems. It is more convenient to solve instead the initial value problems

$$\begin{array}{llll} \mathcal{L} v_1^i = 0 \,, & x_{i-1} \leqq x \leqq x_i \,, & v_1^i(x_{i-1}) = 0 \,, & a(v_1^i)'|_{x=x_{i-1}} = 1 \,, \\ \mathcal{L} v_2^i = 0 \,, & x_i \leqq x \leqq x_{i+1} \,, & v_2^i(x_{i+1}) = 0 \,, & a(v_2^i)'|_{x=x_{i+1}} = -1 \end{array}$$

as in [12], p. 204. These problems are uniquely solvable under our assumption on $a(x)$ and $d(x)$. Thus

$$w_i(x) = \begin{cases} \alpha_i v_1^i(x) \,, & x_{i-1} \leqq x \leqq x_i \\ \beta_i v_2^i(x) \,, & x_i \leqq x \leqq x_{i+1} \\ 0 & x \notin [x_{i-1}, x_{i+1}] \,, \end{cases}$$

where α_i and β_i are selected so as to satisfy

$$\alpha_i v_1^i(x_i) = 1 \,, \qquad \beta_i v^i(x_i) - 1 \,.$$

The functions $v_1^i(x)$ and $v_2^i(x)$ will be used now to express a_{ij} and φ_i.

By applying partial integration to $0 = \int_{x_{i-1}}^{x_i} w_i \mathcal{L} w_i \, \mathrm{d}x$, we find

$$\int_{x_{i-1}}^{x_i} [a w_i'^2 - d w_i^2] \, \mathrm{d}x = a w_i'(x_i - 0) \,.$$

Integrating $\mathcal{L} w_i = 0$ between x_{i-1} and x_i, we obtain

$$a w_i'(x_i - 0) = a w_i'(x_{i-1} + 0) - \int_{x_{i-1}}^{x_i} d(x)\, w_i \, \mathrm{d}x \,.$$

Similarly,

$$\int_{x_i}^{x_{i+1}} [a w_i'^2 - d w_i^2] \, \mathrm{d}x = -a w_i'(x_i + 0) = -a w_i'(x_{i+1} - 0) - \int_{x_i}^{x_{i+1}} d w_i \, \mathrm{d}x \,.$$

Therefore,

$$a_{ii} = \int_{x_{i-1}}^{x_{i+1}} [a w_i'^2 - d w_i^2] \, \mathrm{d}x = a w_i'(x_{i-1} + 0) - a w_i'(x_{i+1} - 0) - \int_{x_{i-1}}^{x_{i+1}} d w_i \, \mathrm{d}x \,,$$

that is,

$$a_{ii} = \alpha_i + \beta_i - V_i(d) \,,$$

where

$$V_i(g) := \int_{x_{i-1}}^{x_i} \alpha_i v_1^i(x)\, g(x)\, \mathrm{d}x + \int_{x_i}^{x_{i+1}} \beta_i v_2^i(x)\, g(x)\, \mathrm{d}x \,.$$

Since, obviously, $a_{ij} = 0$ for $|i - j| > 1$, it remains to reexpress $a_{ii\pm1}$ in terms of v_1^i, v_2^i, α_i and β_i:

$$a_{ii-1} = \int_{x_{i-1}}^{x_i} [a w_i' w_{i-1}' - d w_i w_{i-1}] \, \mathrm{d}x = -a w_i'(x_{i-1} + 0) = -\alpha_i \,,$$

which is obtained by partially integrating $0 = \int_{x_{i-1}}^{x_i} w_{i-1} \mathcal{L} w_i \, \mathrm{d}x$. Analogously,

$$a_{ii+1} = a w_i'(x_{i+1} - 0) = -\beta_i \,.$$

Finally,

$$\varphi_i = V_i(f) \,.$$

Defining $y_0 := 0$, $y_N := 0$, we may write now

$$(Ay)_i = -[\beta_i(y_{i+1} - y_i) - \alpha_i(y_i - y_{i-1}) + y_i V_i(d)] .$$

Remembering that

$$\alpha_i = 1/v_1^i(x_i) , \qquad \beta_i = 1/v_2^i(x_i) ,$$

the Ritz equations (27) are seen to be, up to multiplication by $-2/(x_{i+1} - x_{i-1})$, the equations of the exact difference scheme [12], p. 206.

The consequence that y_i are the values of the solution $u(x)$ of (22) at the knots x_i has been observed for $\mathcal{L}u = u''$ in [21] and for selfadjoint differential operators of order $2n$ in [22] and [18]. Concerning the exact difference scheme, we have the by-product that the difference equations are solvable under the conditions (23), whereas in [11], [12] $a(x) > 0$, $d(x) \leqq 0$ was assumed. Also, the conservativity of the scheme (i.e., the symmetry of the corresponding matrix) follows without calculation from the selfadjointness of the operator $\mathcal{L}$ (and implies $\beta_i = \alpha_{i+1}$).

10. The mildly nonlinear case

To apply the Ritz method, the boundary value problem

$$\mathcal{L}u + f(x, u) = 0 , \qquad 0 < x < 1 , \qquad u(0) = u(1) = 0 , \qquad \mathcal{L}u := au'' + bu' , \tag{28}$$

is first transformed into a selfadjoint one by multiplication with $\exp(bx/a) = e^{2px}$, as in Point 3:

$$\begin{aligned} &-L^*Lu + e^{2px} f(x, u) = 0 , \qquad 0 < x < 1 , \qquad u(0) = u(1) = 0 , \\ &Lu := a^{1/2} e^{px} u' , \qquad L^*u := -a^{1/2}(e^{px} u)' , \end{aligned} \tag{29}$$

and here the nonlinear source term $f(x, u)$ is now assumed to be continuous in $[0, 1] \times (-\infty, \infty)$ and to satisfy the following conditions which guarantee uniqueness both of the solution u of (29) and of the solution y of (27), see [18]:

$$\frac{f(x, u_1) - f(x, u_2)}{u_1 - u_2} \leqq \gamma < \lambda_1 \quad \text{for some} \quad \gamma \geqq 0 , \quad \text{and} \quad 0 \leqq x \leqq 1 , \qquad u_1 \neq u_2 , \tag{30}$$

where

$$\lambda_1 := \min_{\substack{v \in S \\ v \not\equiv 0}} \left\{ \int_0^1 a\, e^{2px} v'^2 \, dx \Big/ \int_0^1 e^{2px} v^2 \, dx \right\} = a\pi^2 + \frac{b^2}{4a} , \tag{30'}$$

further, there exists a constant $M_1(c)$ for every positive number c such that

$$\frac{f(x, u_1) - f(x, u_2)}{u_1 - u_2} \geqq -M_1(c) , \qquad 0 \leqq x \leqq 1 , \qquad |u_1|, |u_2| \leqq c , \qquad u_1 \neq u_2 . \tag{31}$$

We have now, from (25) for $\mathcal{L} = L^*L$,

$$w_i(x) = \begin{cases} (2q)^{-1} [(q - \varrho) \exp(-2p(x - x_i)) + q + \varrho] , & x_{i-1} \leqq x \leqq x_i \\ (2q)^{-1} [(q + \varrho) \exp(-2p(x - x_i)) + q - \varrho] , & x_i \leqq x \leqq x_{i+1} \\ 0 & x \notin [x_{i-1}, x_{i+1}] \end{cases} \tag{32}$$

where $\varrho = \varrho(q) = q \coth q$ as in Point 2.

These basis functions are well-known from finite element context, see, e.g., in [23], the "test functions" $\Psi_i(x)$ on p. 223. In the case $b > 0$ considered further on, it follows that

$$w_i(x) \to 1 , \qquad x_{i-1} < x \leqq x_i , \quad \text{and} \quad w_i(x) \to 0 \quad \text{elsewhere} ,$$

if $a \to 0$.

Now let be $\tilde{w}(x)$ the $\mathcal{L}$-spline interpolating the solution u of (29),

$$\tilde{w} := w(u) = \sum_1^{N-1} u(x_i)\, w_i(x) . \tag{33}$$

Then, from (3.9) and (3.10) in [19], p. 186 (as compared to [19], we write x instead of t, $\varphi(t)$ is changed to $u(x)$, $f(t, u)$ to $-f(x, u)$, $p(t)$ to $r(x)$, and $q(t)$ to $s(x)$):

$$\int_0^1 \{(L[\tilde{w} - \hat{w}])^2 - r(x)\,(\tilde{w} - \hat{w})^2\}\, dx = \int_0^1 s(x)\,(u - \tilde{w})\,(\tilde{w} - \hat{w})\, dx , \tag{34}$$

where now

$$\begin{aligned} &r(x)\,(\tilde{w}(x) - \hat{w}(x)) = e^{2px}[f(x, \tilde{w}(x)) - f(x, \hat{w}(x))] , \qquad \tilde{w}(x) \neq \hat{w}(x) , \\ &r(x) = \gamma\, e^{2px} , \qquad \tilde{w}(x) = \hat{w}(x) . \end{aligned}$$

The function $s(x)$ is defined analogously, with $\hat{w}(x)$ replaced by $u(x)$. The assumptions (30), (31) on f lead to

$$r(x), s(x) \leqq \gamma\, e^{2px} , \qquad |r(x)|, |s(x)| \leqq \Gamma \cdot e^{2px} , \qquad \Gamma := \max(M_1(c_0), \gamma) , \tag{35}$$

and here c_0 is an a priori bound on $||u||_C$, $||\hat{w}||_C$ and $||\tilde{w}||_C$, a problem to which we return below in Points 11, 12 and 14.

Starting from the identity (34), a number of estimates has been established in [19] where the origin of all constants is indicated. Hence, it seems to be sufficient to obtain explicitly those constants and to restate the estimates. This will be done in the point following now.

11. Estimates for the mildly nonlinear case, I

Using (30′), (35) and the definition (29) of the operator L, the left and right-hand sides of (34) are estimated from below and above, respectively:

$$(1-\gamma/\lambda_1)\int_0^1 a\,e^{2px}(\tilde{w}-\hat{w})'^2\,dx \leqq \int_0^1 \{(L[\tilde{w}-\hat{w}])^2 - r(x)\,(\tilde{w}-\hat{w})^2\}\,dx\,,$$

$$\int_0^1 s(x)\,(u-\tilde{w})\,(\tilde{w}-\hat{w})\,dx \leqq \Gamma\left(\int_0^1 e^{2px}(u-\tilde{w})^2\,dx \int_0^1 e^{2px}(\tilde{w}-\hat{w})\,dx\right)^{1/2},$$

$$\leqq \Gamma\lambda_1\left(\int_0^1 e^{2px}(u-\tilde{w})^2\,dx \int_0^1 a\,e^{2px}(\tilde{w}-\hat{w})'^2\,dx\right)^{1/2}.$$

Hence,

$$\left(\int_0^1 a\,e^{2px}(\tilde{w}-\hat{w})'^2\,dx\right)^{1/2} \leqq \frac{\Gamma\lambda_1}{1-\gamma/\lambda_1}\left(\int_0^1 e^{2px}\,(u-\tilde{w})^2\,dx\right)^{1/2}, \qquad (36)$$

which corresponds to (3.14) in [19]. The left and right-hand side integrals can be bounded from below by $K_0\|\tilde{w}-\hat{w}\|_C$ and from above by $K_1\left(\int_0^1 e^{2px} f^2(x,u)\,dx\right)^{1/2}$, respectively. Here, and this is the trouble, neither K_0 nor K_1 contain a factor compensating for the weight e^{2px}.

Lemma 4: *For $v \in S$, there holds*

$$\|v\|_C \leqq K_0\left(\int_0^1 a\,e^{2px}\,v'^2\,dx\right)^{1/2},$$

$$K_0 := \left(\frac{1}{4b}\,(1-e^{-2p})\right)^{1/2} \leqq (4\max(a,b))^{-1/2}, \qquad p = b/2a\,,$$

and, for $p \geqq 0.8$, there exists a function $\bar{v} = \bar{v}(x,a,b) \in S$ with

$$\|\bar{v}\|_C = K_0\left(\int_0^1 a\,e^{2px}\,\bar{v}'^2\,dx\right)^{1/2},\ \bar{K}_0 = (b\cdot 6.176\ldots)^{-1/2}.$$

Proof: In the manner of [12], p. 119, we write

$$|v(x)| \leqq \left|\int_0^x v'\,dx\right| \leqq \left(\int_0^x v'^2 a\,e^{2px}\,dx \int_0^x \frac{1}{a}\,e^{-2px}\,dx\right)^{1/2},$$

that is,

$$v^2(x) \leqq \frac{1}{b}\,(1-e^{-2px})\int_0^x a\,e^{2px}\,v'^2\,dx\,.$$

Similarly,

$$v^2(x) \leqq \frac{1}{b}\,(e^{-2px}-e^{-2p})\int_x^1 a\,e^{2px}\,v'^2\,dx\,.$$

Multiplying these inequalities by $(e^{-2px}-e^{-2p})$ and $(1-e^{-2px})$, respectively, and adding, we obtain

$$(1-e^{-2p})\,v^2(x) \leqq \frac{1}{b}\,g(x)\int_0^1 a\,e^{2px}\,v'^2\,dx\,,$$

where $g(x) := (1-e^{-2px})\,(e^{-2px}-e^{-2p})$. The function $g(x)$ takes on its maximum at x_0 with $e^{2px_0} = 2e^{2p}/(e^{2p}+1)$. Therefore,

$$0 \leqq g(x) \leqq \left(\frac{1}{2}\,(1-e^{-2px})\right)^2 \quad\text{and}\quad (b(1-e^{-2p}))^{-1}\,g(x) \leqq \frac{1}{4b}\,(1-e^{-2p}) \leqq \frac{1}{4b}\min\left(\frac{b}{a},1\right),$$

and the inequality of the lemma follows.

To prove the second part of the assertion, for $\frac{1}{2} > \delta > 0$ consider

$$\bar{v} = \begin{cases} x/\delta\,, & 0 \leqq x \leqq \delta \\ 2 - x/\delta\,, & \delta \leqq x \leqq 2\delta \\ 0\,, & 2\delta \leqq x \leqq 1\,. \end{cases}$$

Then $\bar{v} \in S$, $\|\bar{v}\|_C = 1$ and $\int_0^1 a\,e^{2px}\,\bar{v}'^2\,dx = \frac{a}{2p\delta^2}\,(e^{4p\delta}-1) =: E(\delta)$. The function $E(\delta)$ is minimal at δ_0 with $p\delta_0 = r_0$, $e^{2r_0} = 1/(1-r_0)$, which gives $r_0 = 0.7968\ldots$ and $E(\delta_0) = b/r_0(1-r_0) = b\cdot 6.176\ldots$

Lemma 5: *For u and $\tilde{w}$ defined by* (29) *and* (33), *there holds*

$$\left(\int_0^1 e^{2px}(u-\tilde{w})^2\,dx\right)^{1/2} \leqq \frac{h^2}{a\pi^2+h^2b^2/4a}\left(\int_0^1 e^{2px} f^2(x,u)\,dx\right)^{1/2}.$$

Proof: Using $u - \tilde{w} = 0$ at $x = x_i$, $i = 0, \dots, N$, we have firstly, by partial integration,

$$\int_0^1 a\, e^{2px}(u-\tilde{w})'^2\,dx = \int_0^1 e^{2px} f(x,u)\,(u-\tilde{w})\,dx\,,$$

comp. (3.4) in [20], p. 351. Therefore,

$$\int_0^1 a\, e^{2px}(u-\tilde{w})'^2\,dx \leqq \Big(\int_0^1 e^{2px} f^2\,dx \int_0^1 e^{2px}(u-\tilde{w})^2\,dx\Big)^{1/2}. \tag{37}$$

Secondly, from (30′) for $w(x) \in S$ satisfying $w(x_i) = 0$, $i = 0, \dots, N$,

$$\int_{x_{i-1}}^{x_i} \exp\left(\frac{bh}{a}\,\frac{(x-x_{i-1})}{h}\right) w^2\,dx \leqq \frac{h^2}{a\pi^2+h^2b^2/4a}\int_{x_{i-1}}^{x_i} a\exp\left(\frac{bh}{a}\,\frac{(x-x_{i-1})}{h}\right) w'^2\,dx\,.$$

Multiply this by $\exp\left(\frac{b}{a}\,x_{i-1}\right)$, add over i and substitute $u - \tilde{w}$ for w. Then

$$\int_0^1 e^{2px}\,(u-\tilde{w})^2\,dx \leqq \frac{h^2}{a\pi^2+h^2b^2/4a}\int_0^1 a\, e^{2px}(u-\tilde{w})'^2\,dx$$

results. This inequality and (37) combine to give the assertion of the lemma.

We remark that, elementarily, $h^2/(a\pi^2 + h^2b^2/4a) \leqq \frac{h}{\pi}\min\left(\frac{1}{b}, \frac{h}{a\pi}\right)$. Collecting the inequalities of the Lemmas 4, 5 and of (36), we obtain

$$\|\tilde{w} - \hat{w}\|_C \leqq \frac{\Gamma\lambda_1 h \min\left(\frac{1}{b}, \frac{h}{a\pi}\right)}{(1-\gamma/\lambda_1)\,(4\max(a,b))^{1/2}\,\pi}\left(\int_0^1 e^{2px} f^2(x,u)\,dx\right)^{1/2}.$$

This then constitutes the equivalent to (3.15) in [19], p. 186, with explicitly given constants.

From here, using the triangle inequality, a maximum norm estimate of $u - \hat{w}$ could be obtained. But already here the estimate is seen to become useless in case $a \to 0$, i.e., $p \to \infty$.

To finish this point, we state a maximum norm estimate for $u(x)$ which is derived following [24], p. 72, and employing Lemma 4:

$$\|u\|_C \leqq M_0 \sinh p/\big((1-\gamma/\lambda_1)\, b(\pi^2+p^2)\big)^{1/2} \tag{38}$$

where $M_0 := \max_{[0,1]} |f(x, 0)|$, a constant which we shall need below repeatedly. The estimate, like that of $\|\tilde{w} - \hat{w}\|_C$, becomes useless for $a \to 0$.

12. Estimates for the mildly nonlinear case, II

In the preceding point, using integral inequalities and partial integration, we did not succeed in dropping the weighting function e^{2px} introduced to have a selfadjoint problem. Therefore, we shall first cancel this factor and then estimate.

According to Point 9, the Ritz equations (27) multiplied by $-h^{-1}$ (since $2/(x_{i+1} - x_{i-1}) = h$ on the equidistant grid) are just the equations of the exact difference scheme. We remember also that the formulae (3)—(7) of Point 3 had been obtained after cancelling a common factor e^{2px} from the equations of the exact difference scheme. Thus, multiplying the i-th Ritz equation (27) by $-(he^{2px_i})^{-1}$, we return, in fact, to the difference scheme (3)—(7). That is, we consider now the nonlinear difference scheme (3)—(7), $f = f(x, w(y))$, where $w(y) = \hat{w}$ is the $\mathscr{L}$-spline (26) interpolating the solution y by means of the functions (32).

Returning in principle from [19] to [10], we shall estimate the accuracy of the nonlinear scheme using the maximum principle, i.e., the nonnegativity of the Green's function.

Lemma 6: *Assume $f(x, u)$ to be continuous in* $[0, 1] \times (-\infty, \infty)$ *and to satisfy, for $\gamma \geqq 0$,*

$$\frac{f(x,u_1)-f(x,u_2)}{u_1-u_2} \leqq \gamma < \frac{1}{\|Y\|_C}, \qquad 0 \leqq x \leqq 1\,, \qquad u_1 \neq u_2\,. \tag{39}$$

Then for the solution u of (28) *and with Y from* (13), (14), *the following estimate holds*:

$$\|u\|_C \leqq M_0 \|Y\|_C (1 - \gamma \|Y\|_C)$$

Proof: We have

$$\mathcal{L}u + ru = -f(x, 0), \qquad 0 < x < 1, \qquad u(0) = u(1) = 0,$$

where $ru := f(x, u) - f(x, 0)$, $u \neq 0$, $r := \gamma$, $u = 0$. From (39), $r \leqq \gamma$. Therefore,

$$\mathcal{L}(mY \pm u) + r(mY \pm u) = -m + rmY \mp f(x, 0) \leqq -m(1 - \gamma \|Y\|_C) + M_0 = 0,$$

if $m = M_0/(1 - \gamma \|Y\|_C)$ is selected. This inequality implies $mY \pm u \geqq 0$ and, hence, the estimate of the lemma, if the Green's function of the operator $-(\mathcal{L} + r)$ is nonnegative.

By (39), and remembering the Remark to Theorem 1, $r < \lambda_1$. Therefore, the nonnegativity of the Green's function $G(x, \xi)$ of the selfadjoint operator $L^*L - e^{2px} r$ (with L^*L defined by (29)) follows from [25] and involves nonnegativity of $G(x, \xi)\, e^{2p\xi}$, the Green's function of $-(\mathcal{L} + r)$.

Remark: The same bound applies to $\|\tilde{w}\|_C$. For $x \in [x_{i-1}, x_i]$, $\tilde{w}$ is characterized by

$$\mathcal{L}\tilde{w} = 0, \qquad x_{i-1} < x < x_i, \qquad \tilde{w}(x_j) = u(x_j), \qquad j = i - 1, i.$$

The maximum principle then gives

$$\min\big(u(x_i), u(x_{i-1})\big) \leqq \tilde{w}(x) \leqq \max\big(u(x_i), u(x_{i-1})\big), \qquad x_{i-1} \leqq x \leqq x_i,$$

which means

$$\|\tilde{w}\|_C = \|u\|_{C,h} \leqq \|u\|_C.$$

To be sure of the nonnegativity of the Green's function of the discrete problem, we need a strengthened form of (31) and a restriction on h.

Lemma 7: *Assume f to be continuous and to satisfy*

$$-M_1 \leqq \frac{f(x, u_1) - f(x, u_2)}{u_1 - u_2} \leqq \gamma \,\hat{<}\, \frac{1}{\|Y\|_{C,h}}, \qquad 0 \leqq x \leqq 1, \qquad u_1 \neq u_2, \tag{40}$$

where M_1 and γ are fixed nonnegative numbers. Moreover, let the steplength h be restricted by

$$h \leqq \max\left(\left(\frac{6a}{M_1}\right)^{1/2}, \frac{b}{M_1}\right). \tag{41}$$

Then for the solution of (3)—(7), $f = f(x, w(y))$, *the following estimate holds*

$$\|y\|_{C,h} \leqq M_0 \|Y\|_{C,h} / (1 - \gamma \|Y\|_{C,h}).$$

Proof: Analogously as in the proof of Lemma 6 we have

$$\Lambda y + \varphi(r\hat{w}) = -\varphi\big(f(x, 0)\big), \qquad x \in \omega_h, \qquad y_0 = y_N = 0. \tag{42}$$

From (6) and (7) — or employing $\varphi_i(f) = (h\, e^{2px_i})^{-1} \int_0^1 e^{2px} w_i(x) f(x)\, dx$, comp. Point 9 and the beginning of this point —

$$\varphi_i(w_{i\pm1}) = (\varrho \pm q)\, \frac{\alpha}{q}, \qquad \varphi_i(w_i) = 1 - 2\varrho \frac{\alpha}{q}, \qquad \varphi_i(w_j) = 0, \qquad |i - j| > 1.$$

All these quantities are nonnegative, see the Appendix.

Using the mean value theorem, we may write

$$\varphi_i(r\hat{w}) = (\Phi_R y)_i, \qquad \Phi_R := (r^*_{ij} \Phi_{ij}), \qquad \Phi = (\Phi_{ij}) := \big(\varphi_i(w_j)\big), \tag{43}$$

where $-M_1 \leqq r^*_{ij} \leqq \gamma$, from (40), the star denoting some mean value.

Now let be A the matrix corresponding to the difference operator $-(\Lambda + \Phi_R)$. We use the well-known criterion that A is an M-matrix (and hence $A^{-1} \geqq 0$) if the off-diagonal elements of A are nonpositive and if there is a majorizing element (a nonnegative vector e satisfying $Ae > 0$), see, e.g., [26], Theorem 10.7 on p. 173. Since $(A_\Lambda)_{ii\pm1} = -(\varrho \pm q)\, a/h^2$ for the matrix A_Λ corresponding to $-\Lambda$, the off-diagonal elements of A are nonpositive if

$$-\frac{a}{h^2} - r^* \frac{\alpha}{q} \leqq 0.$$

This certainly is fulfilled if

$$M_1 \min\left(\frac{1}{6}, \frac{1}{2q}\right) \leqq \frac{a}{h^2}, \tag{44}$$

comp. the Appendix. (44) leads to (41).

Next, we use $r^* \leqq \gamma$ and $0 \leqq Y \leqq \|Y\|_{C,h}$, $x \in \bar{\omega}_h$, to obtain

$$-\big((\Lambda + \Phi_R)\, Y\big)_i = 1 - \left[r^*_{ii-1} Y_{i-1}(\varrho - q) \frac{\alpha}{q} + r^*_{ii} Y_i \left(1 - 2\varrho \frac{\alpha}{q} + r^*_{ii-1} Y_{i+1}(\varrho + q) \frac{\alpha}{q}\right.\right]$$

$$\geqq 1 - \gamma \|Y\|_{C,h} > 0.$$

Now we know the Green's function of $-(\Lambda + \Phi_R)$ to be nonnegative. We have $|\varphi_i(f(x, 0))| \leqq M_0$ by Lemma 2, and the estimate of $||y||_{C,h}$ follows.

Remarks:

1. If the restriction (41) on h is replaced by the restriction

$$M_1 \leqq 2 \max (6a, b) \tag{45}$$

on M_1, then the lemma remains true. This follows from (44) taking into account that we may assume $h \leqq \frac{1}{2}$.

2. The above considerations imply, under the conditions (41) or (45),

$$||A^{-1}||_C \leqq ||Y||_{C,h}/(1 - \gamma\, ||Y||_{C,h}) .$$

Remembering $||Y||_{C,h} \leqq ||Y||_C \leqq \min\left(\frac{1}{8a}, \frac{1}{b}\right)$ from Lemma 1, we have

$$||A^{-1}||_C \leqq (8 - \gamma)^{-1}$$

in case $a = 1$, $b = 0$, which corresponds just to (2.31) in [10].

3. Applying once more the maximum principle, we find $||\hat{w}||_C = ||y||_{C,h}$ as in the Remark to Lemma 6. Hence the bound of Lemma 6 is valid for $||u||_C$, $||\tilde{w}||_C$ and, if (41) or (45) hold true, for $||\hat{w}||_C$, too.

We are now in position to prove the following accuracy estimate.

Theorem 4: *Let be $u(x)$ the solution of (28), (40) and y the solution of the difference scheme (3)–(7), $f = f(x, w(y))$, with $w(y)$ defined by (26) and (32). Then, if $\gamma < \max (8a, b)$ and either (41) or (45) are satisfied, the following inequality holds:*

$$||u - y||_{C,h} \leqq \frac{\Gamma M(c_0)\, C_Y}{1 - \gamma C_Y} h \min\left(\frac{h}{6a}, \frac{1}{b}\right),$$

where

$$\Gamma := \max (M_1, \gamma) , \qquad C_Y := \min\left(\frac{1}{8a}, \frac{1}{b}\right),$$

$$M(c_0) := \max_{\substack{[0,1]\\ |v| \leqq c_0}} |f(x, v)|, \qquad c_0 := \frac{1}{1 - \gamma C_Y} M(0)\, C_Y .$$

Proof: Since the scheme is exact in the linear case, we have

$$\Lambda u + \varphi\big(f(x, u(x))\big) = 0 , \qquad x \in \omega_h , \qquad u(0) = u(1) = 0 ,$$

for the solution of (28). Moreover, $\Lambda u = \Lambda\tilde{w}$, because of $u(x) = \tilde{w}(x)$, $x \in \omega_h$, for the spline function (33). Thus, the spline function z,

$$z := \tilde{w} - \hat{w} = \sum_1^{N-1} \big(u(x_i) - y_i\big)\, w_i(x) ,$$

satisfies

$$\Lambda z + \varphi(rz) = \psi , \qquad x \in \omega_h , \qquad z_0 = z_N = 0 , \qquad \psi := \varphi\big(s(\tilde{w} - u)\big) ,$$

where

$$r(x)\,\big(\tilde{w}(x) - \hat{w}(x)\big) = f\big(x, \tilde{w}(x)\big) - f\big(x, \hat{w}(x)\big) , \qquad \tilde{w}(x) \neq \hat{w}(x) , \qquad r(x) = \gamma , \qquad \tilde{w}(x) = \hat{w}(x) ,$$
$$s(x)\,\big(\tilde{w}(x) - u(x)\big) = f\big(x, \tilde{w}(x)\big) - f\big(x, u(x)\big) , \qquad \tilde{w}(x) = u(x) , \qquad s(x) = \gamma , \qquad \tilde{w}(x) = u(x) .$$

By assumption,

$$-M_1 \leqq r, s \leqq \gamma < 1/\,||Y||_{C,h} .$$

We use the notation (43) where, again, $-M_1 \leqq r_{ij}^* \leqq \gamma$. We have then that the Green's function of $-(\Lambda + \Phi_R)$ is nonnegative and

$$||u||_C,\ ||\tilde{w}||_C,\ ||\hat{w}||_C \leqq c_0 ,$$

and furthermore, $||z||_{C,h} \leqq C_Y/(1 - \gamma C_Y)\, ||\psi||_{C,h}$, see Lemma 7 and the remarks to that lemma. It remains to estimate $||\psi||_{C,h}$. Applying Lemma 2, we get

$$|\psi_i| \leqq \frac{\Gamma}{h} \int_{x_{i-1}}^{x_{i+1}} |\tilde{w} - u| \, dx ,$$

and the integral is estimated by the maximum principle as follows:

The function $\tilde{w}(x) - u(x)$ solves

$$\mathcal{L}(\tilde{w} - u) + f(x, u) = 0 , \qquad x_{i-1} < x < x_i , \qquad \tilde{w}(x_j) - u(x_j) = 0 , \qquad j = i - 1, i .$$

Therefore

$$|\tilde{w}(x) - u(x)| \leqq v(x) , \qquad x_{i-1} \leqq x \leqq x_i ,$$

where $v(x)$ is the solution of

$$\mathcal{L} v + M(c_0) = 0 , \qquad x_{i-1} < x < x_i , \qquad v(x_{i-1}) = v(x_i) = 0 .$$

Solving this boundary value problem exactly, we find

$$v(x) = \frac{M(c_0)}{b} h \left[\frac{\varrho + q}{2q} (1 - e^{-2p(x - x_{i-1})} - \frac{x - x_{i-1}}{h} \right],$$

and

$$\int_{x_{i-1}}^{x_i} v(x)\, dx = \frac{M(c_0)}{b} h^2 \alpha(q) .$$

Here the right-hand side is independent on i. Consequently,

$$\int_{x_{i-1}}^{x_i} |\tilde{w} - \dot{u}|\, dx \leqq \frac{M(c_0)}{b} h^2 \alpha(q) , \qquad i = 1, \dots, N ,$$

and

$$||\psi||_{C,h} \leqq 2 \frac{\Gamma M(c_0)}{b} h\alpha(q) \leqq \Gamma M(c_0)\, h \min\left(\frac{h}{6a}, \frac{1}{b}\right),$$

see the Appendix, and the proof is finished.

Remarks:

1. By further use of the maximum principle, an estimate of $||u - \hat{w}||_C$ is readily obtained: For the above function $v(x)$ majorizing $|u(x) - \tilde{w}(x)|$ we have, elementarily,

$$||v||_C \leqq M(c_0)\, h \min\left(\frac{h}{2a}, \frac{1}{b}\right).$$

Moreover, applying the maximum principle to

$$\mathcal{L}(\tilde{w} - \hat{w}) = 0 , \qquad x_{i-1} < x < x_i ,$$

with $\tilde{w} - \hat{w}$ prescribed at x_{i-1} and x_i, we find

$$||\tilde{w} - \hat{w}||_C = ||\tilde{w} - \hat{w}||_{C,h} = ||u - \hat{w}||_{C,h} = ||u - y||_{C,h} .$$

Using this and the triangle inequality, we get

$$||u - \hat{w}||_C \leqq ||u - y||_{C,h} + ||v||_C .$$

2. It may be convenient to replace the basis functions (32) by simpler ones. For example, let be $\bar{y}$ the solution of (3)–(7), $f = f(x, \overline{w}(\bar{y}))$,

$$\overline{w}(\bar{y}) := \sum_1^{N-1} \bar{y}_i \overline{w}_i(x) ,$$

where $\overline{w}_i(x)$ are the stepfunctions to which the $w_i(x)$ converge in case $b > 0$ $a \to 0$, comp. after (32).

A similar analysis as in the proof of Theorem 4 shows that the additional error caused by the replacement of the w_i by $\overline{w}_i$ is bounded by

$$\frac{2}{q} \Gamma M_0 \left(\frac{||Y||_C}{1 - \gamma\, ||Y||_C} \right)^2 ,$$

which is acceptable only for $\frac{1}{q} \leqq c_1 h$, i.e., $a \leqq \frac{1}{2} c_1 bh^2$, for some positive constant c_1.

13. On the solution of the nonlinear difference scheme

An effective solution of the system of nonlinear equations represented by (3)–(7), $f = f(x, w(y))$, is achieved by multi-grid methods, see e.g., [27] and [28]. The linear equations arising in the iteration are solved by the usual shortened Gauß elimination [12], p. 40.

Since it is rather instructive, we shall shortly consider two simple iteration methods on the fixed grid ω_h.

As is well-known, the boundary value problem

$$\mathcal{L}u + f(x, u) = 0 , \qquad 0 < x < 1 , \qquad u(0) = u(1) = 0 ,$$

can be solved by the iteration process

$$u^0 = 0 , \qquad \mathcal{L}u^n + f(x, u^{n-1}) = 0 , \qquad 0 < x < 1 , \qquad u(0) = u(1) = 0 , \qquad n = 1, 2, \dots , \tag{46}$$

if $f(x, 0) \geqq 0$ and $f(x, u) \geqq f(x, v)$ for $u \geqq v$, see [29] or [30], p. 46. Likewise, the iteration

$$u^0 = 0 , \qquad (\mathcal{L} - M_1)\, u^n + f(x, u^{n-1}) + M_1 u^{n-1} = 0 , \qquad 0 < x < 1 , \qquad u(0) = u(1) = 0 , \qquad n = 1, \dots$$

may be used (M_1 being the constant in (40)) if $\frac{\partial f}{\partial u}$ exists and $\frac{\partial}{\partial u} (f(x, u)/u) < 0$, see [31] and also [30], p. 44. Clearly, the conditions on f must be satisfied only for $0 \leqq x \leqq 1$, $|u| \leqq c_0$, c_0 the constant from Theorem 4.

Both iterations give rise to monotonically increasing sequences $\{u^n\}$, which are convergent since the iterates are uniformly bounded by c_0. Since all considerations rest here on the nonnegativity of the Green's functions of $-\mathcal{L}$ and $-\mathcal{L} + M_1$, the convergence results immediately carry over to the difference scheme (3)–(7), $f = f(x, w(y))$. For example, the discrete formulation of the second iteration process is

$$y^0 = 0\,, \qquad (\Lambda - M_1\Phi)\, y^n + \varphi(f(x, w(y^{n-1})) + M_1\Phi y^{n-1} = 0\,, \qquad x \in \omega_h\,,$$
$$y_0^n = y_N^n = 0\,, \qquad n = 1, 2, \ldots\,,$$

where Φ is defined by (43), and (41) or (45) must be satisfied.

In the above monotone iterations, f is assumed to fullfil one-sided conditions. To finish this point we shall prove that the discretized version of (46) converges also under a two-sided condition.

Theorem 5: *Suppose that $f(x, u)$ is continuous and satisfies (40) with $M_1 = \gamma < \max(8a, b)$. Then the iteration*

$$y^0 = 0\,, \qquad \Lambda y^n + \varphi(f(x, w(y^{n-1})) = 0\,, \qquad x \in \omega_h\,, \qquad y_0^n = y_N^n = 0\,, \qquad n = 1, \ldots\,, \tag{46_h}$$

converges (with the speed μ^n where $0 \leqq \mu < 1$ and μ does not depend on h) to the unique solution of (3)–(7), $f = f(x, w(y))$.

Proof: We show that the grid function Nemytzki operator T,

$$(Ty)_i := \sum_1^{N-1} G_{ij}^h \varphi_j(f(x, w(y)))\, h\,, \qquad i = 1, \ldots, N-1\,, \qquad (Ty)_0 := 0\,, \qquad (Ty)_N := 0\,,$$

is a contraction in a norm equivalent to $||\cdot||_{C,h}$. Here G^h is the Green's function (11). Using the notation (43), we may write

$$(Ty^1 - Ty^2)_i = \sum_1^{N-1} G_{ij}^h \varphi_j(r[w(y^1) - w(y^2)])\, h = \sum_1^{N-1} G_{ij}^h (\Phi_R(y^1 - y^2))_j\, h\,,$$

where Φ_R is defined analogously as in the proof of Lemma 7, and $|r_{ij}^*| \leqq \gamma$. By the nonnegativity of G^h and of Φ (for the latter, see also the proof of Lemma 7),

$$|(Ty^1 - Ty^2)_i| \leqq \gamma \sum_1^{N-1} G_{ij}^h (\Phi|y^1 - y^2|)_j\, h\,. \tag{47}$$

We introduce now the auxiliary function g, the solution of

$$\Lambda g + \gamma\Phi g + \gamma = 0\,, \qquad x \in \omega_h\,, \qquad g_0 = g_N = 0\,. \tag{48}$$

Compare now (43) and Point 9 to see that the equations (48) are quite close to those of the exact difference scheme to

$$\mathcal{L}g + \gamma g + \gamma = 0\,, \qquad 0 < x < 1\,, \qquad g(0) = g(1) = 0\,, \tag{49}$$

the only difference being that in (48) the basis functions (32) have been used which correspond to $\mathcal{L}$ and not to $\mathcal{L} + \gamma$.

From the Remark to Theorem 1 we know $\gamma < \lambda_1$. Hence (48) defines a unique solution g. By the nonnegativity of the Green's function of $-(\Lambda + \gamma\Phi)$, valid under our assumptions without restrictions on h (see Remark 1 to Lemma 7), g is nonnegative. Therefore, a norm equivalent to $||\cdot||_{C,h}$ is introduced by

$$||y||_{g,h} := \max_{x\in\omega_h} \frac{|y(x)|}{1 + g(x)}\,, \tag{50}$$

for grid functions y satisfying $y_0 = y_N = 0$.

Using

$$(\Phi|y^1 - y^2|)_j \leqq ||y^1 - y^2||_{g,h}\, (\Phi(1 + g))_j\,,$$

we obtain from (47)

$$|(Ty^1 - Ty^2)_i| \leqq ||y^1 - y^2||_{g,h}\, \gamma \sum_1^{N-1} G_{ij}^h (\Phi(1 + g))_j\, h\,.$$

From the definition of g and G^h it follows that

$$g_i = \gamma \sum_1^{N-1} G_{ij}^h (\Phi(1 + g))_j\, h\,,$$

which implies

$$|(Ty^1 - Ty^2)_i| \leqq ||y^1 - y^2||_{g,h}\, g_i\,.$$

After division by $1 + g_i$, this leads to

$$||Ty^1 - Ty^2||_{g,h} \leqq \mu\, ||y^1 - y^2||_{g,h}\,,$$

where

$$\mu := \max_{x\in\omega_h} \frac{g(x)}{1 + g(x)} = 1 - 1/(1 + ||g||_{C,h})\,.$$

Here μ is seen to be less than 1 and bounded away from 1. Using Lemma 7 we get

$$||g||_{C,h} \leqq \gamma ||Y||_{C,h}\, (1 - \gamma\, ||Y||_{C,h})^{-1} \quad \text{and} \quad \mu \leqq \gamma ||Y||_{C,h} \leqq \gamma \max(8a, b) < 1\,.$$

Thus, the iteration $y^0 = 0$, $y^n = Ty^{n-1}$, $n = 1, \ldots$, or, equivalently, (46_h), converges as asserted (and, in fact, for any y^0 satisfying $||y^0||_{C,h} \leqq c_0$), see, e.g. [30], p. 25.

Remarks:

1. The above approach works also for the continuous problem and unter general conditions assuring the Green's function to be nonnegative. The norm (50) is constructed after that of [30], p. 12, for an ordinary differential equation (i.e., $a = 0$, $b = -1$ in our context) with Lipschitz continuous right-hand side and for the Lipschitz constant γ, the idea originating from [32].

2. In the case $b = 0$, the result of Theorem 5 has been obtained in another way in [33], p. 273.

14. Once more on maximum norm estimates

As is well-known, the contraction principle may also be used to obtain an a priori estimate of $||y||_{g,h}$ and hence of $||y||_{C,h}$, or, in the continuous case, of $||u||_C$. But a better estimate is found by direct use of the auxiliary function $g(x)$ as majorizing function. On the grid ω_h, $g(x)$ is defined by (48), and on the interval [0, 1], by (49).

In a quite analogical way as in the proofs of Lemmas 6 and 7 we find the bound

$$||y||_{C,h},\ ||u||_C \leqq M_0\gamma^{-1}||g||_C\,, \qquad M_0 = \max_{[0,1]} |f(x, 0)|\,,$$

if

$$-2\max(6a, b) \leqq \frac{f(x, u_1) - f(x, u_2)}{u_1 - u_2} \leqq \gamma < \frac{1}{||Y||_{C,h}}, \qquad 0 \leqq x \leqq 1\,, \qquad u_1 \neq u_2\,,$$

is true. (It looks even more suited for estimating $||u||_C$ to take, as majorizing function, the solution of

$$\mathcal{L}v + \gamma v + M_0 = 0\,, \qquad 0 < x < 1\,, \qquad v(0) = v(1) = 0\,.$$

But $v = M_0\gamma^{-1}g$, and the final estimate is the same.)

We give three examples of special solutions of (49) together with $||g||_C$ as needed for the maximum norm bound on $u(x)$, and with contraction constant $\mu = 1 - 1/(1 + ||g||_C)$:

a) $a = 1$, $b = 0$, $\gamma < \pi^2$,

$$s := \sqrt{\gamma}\,, \qquad g = \cos sx - 1 + (1 - \cos s)\sin sx/\sin s\,,$$

$$||g||_C = 2\sin^2\frac{s}{4}\Big/\cos\frac{s}{2}\,, \qquad \mu = 1 - \cos\frac{s}{2}\,;$$

b) $a = 0$, $b = -1$, γ any positive constant

$$g = e^{\gamma x} - 1\,, \qquad ||g||_C = e^{\gamma} - 1\,, \qquad \mu = 1 - e^{-\gamma};$$

c) $a, b > 0$, $\gamma = b^2/4a$ (We choose this value for simplicity of calculation. We have then $\gamma < \lambda_1$, $|\gamma - \lambda_1| \to 0$, $a \to 0$.)

$$g = (1 - x + x\,e^p)\,e^{-px} - 1\,,$$

$$x_m := \tfrac{1}{2} - \alpha(p/2)\,, \qquad ||g||_C = (1 - x_m + x_m\,e^p)\,e^{-px_m} - 1\,, \qquad \mu = 1 - (1 - x_m + x_m\,e^p)^{-1}\,e^{px_m}\,.$$

Example b) corresponds just to [32]. Example c) is quite instructive: From the inequalities on the function α in the Appendix below, we have

$$\frac{1}{2 + p} \leqq x_m \leqq \min\left(\frac{1}{2}, \frac{1}{p}\right) \leqq \frac{2}{2 + p},$$

and this shows $||g||_C = O\left(\frac{1}{p}\,e^p\right)$ for great p. In other words, allowing γ to apparoch to λ_1 we obtain an estimate of the type (38) — which is seen to reflect the general situation properly.

We conclude that the nonnegativity of the Green's function is easy to employ and connects the limiting cases ($a > 0$, $b = 0$ and $a = 0$, $b \neq 0$) in a natural way. However, the result of Theorem 4 (which is senseful under the conditions (2)) rests also on suitably bounding γ away from λ_1.

15. Appendix

The auxiliary functions $\varrho(q)$ and $\alpha(q)$ are even and odd in q, respectively, and sign α = sign q = sign b. For $\varrho(q)$, elementarily,

$$\max(1, |q|) \leqq \varrho(q) \leqq \min\left(1 + \frac{q^2}{3}, 1 + \frac{q^2}{1 + |q|}\right).$$

(The right-hand inequalities as well as the stronger estimate

$$\varrho(q) \leqq \left(1 + \frac{2 + q^2}{3 + q^2}\,q^2\right)^{1/2}$$

are of the type $\varrho(q) \leqq s(q)$, $s \geqq \max(1, |q|)$, which is equivalent to $e^{2q} \geqq (s(q) + q)/(s(q) - q)$. Now the inequalities can be proved by replacing e^{2q} by a sufficiently long part of its Taylor series. From below, we may use $\varrho(q) \geqq |q|$ for $|q| \geqq 2$, and $\varrho(q) \geqq 1 + \frac{1}{4}q^2$ for $0 \leqq |q| \leqq 2$, and this estimate can be obtained by employing

$$e^{2q} \leqq \sum_0^n \frac{(2q)^m}{m!} + \frac{(2q)^{n+1}}{(n + 1)!}\,e^{2q}$$

for $n = 8$.)

For small q the following development is found:

$$\varrho(q) = 1 + \tfrac{1}{3} q^2 - \tfrac{1}{45} q^4 + O(q^6) .$$

The above relations imply for $\alpha(q) = (\varrho(q) - 1)/2q$

$$0 \leqq \max\left(0, \frac{|q| - 1}{2|q|}\right) \leqq |\alpha(q)| \leqq \min\left(\frac{|q|}{6}, \frac{|q|}{2(1 + |q|)}\right) \leqq \frac{1}{2},$$

and

$$\alpha(q) = \frac{q}{6} - \frac{q^3}{90} + O(q^5) .$$

Therefore,

$$\lim_{q\to 0} \alpha(q) = 0 , \qquad \lim_{q\to 0} \frac{6\alpha}{q} = 1 , \qquad \lim_{q\to 0} \frac{1}{q}\left(\frac{6\alpha}{q} - 1\right) = 0 ,$$

and

$$\lim_{|q|\to\infty} \alpha(q) = \tfrac{1}{2} \operatorname{sign} b .$$

For the function ζ the following inequalities are true:

$$0 = \zeta(1) \leqq \zeta(h) \leqq \zeta(0) := \lim_{h\to 0} \zeta(h) = \coth p - \frac{1}{p} = 2\alpha(p) , \qquad 0 \leqq h \leqq 1 , \qquad p > 0 .$$

Indeed, the elementary inequality

$$0 \geqq 2ph - \sinh 2ph$$

implies

$$0 \geqq \frac{ph}{\sinh^2 ph} - \coth ph = ph(\coth^2 ph - 1) - \coth ph = \zeta'(h) .$$

Acknowledgement

The author thanks the referee for critical remarks.

References

1 Емельянов, К. В., О разностной схеме для обыкновенного дифференциального уравнения с малым параметром, Ж. вычисл. матем. и матем. физ. **18**, 5 (1978), 1146—1153.

2 Kellog, R. B.; Tsan, A., Analysis of some difference approximations for a singular perturbation problem without turning points, Math. Comput. **32** (1978), 1025—1039.

3 Hemker, P. W.; Miller, J. J. H. (eds.), Numerical Analysis of Singular Perturbation Problems, Academic Press, London 1979.

4 Алексеевский, М. В., О разностной схеме для дифференциального уравнения с малым параметром при старшей производной. В сб.: Разностные методы математической физики. МГУ, Москва 1979.

5 Lorenz, J., Zur Theorie und Numerik von Differenzenverfahren für singuläre Störungen, Habilitationsschrift, Universität Konstanz 1980.

6 Doolan, E. P.; Miller, J. J. H.; Schilders, W. H. A., Uniform Numerical Methods for Problems with Initial and Boundary Layers, Boole Press, Dublin 1980.

7 Алексеевский, М. В., Разностные схемы высокого порядка точности для сингулярно возмущенной краевой задачи, Дифф. уравнения XII, 7 (1981), 1171—1183.

8 Ильин, А. М., Разностная схема для дифференциального уравнения с малым параметром при старшей производной, Матем. зам. **6**, 2 (1969), 237—248.

9 Stoyan, G., Monotone difference schemes for diffusion-convection problems, ZAMM 59 (1979), 361—372.

10 Ciarlet, P. G., An $O(h^2)$ method for a non-smooth boundary value problem, Aequationes math. **2**, 1 (1968), 39—49.

11 Тихонов, А. Н.; Самарский, А. А., Об однородных разностных схемах высокого порядка точности, Доклады АН СССР, **131**, 3 (1960), 514—517.

12 Самарский, А. А., Теория разностных схем, Наука, Москва 1977.

13 Stoyan, G., Über eine monotone Differenzenapproximation einer partiellen Differentialgleichung. In: Report R-05/80, (ed. W. Zwick), ZIMM der AdW der DDR, Berlin 1980, S. 83—94.

14 Штоян К асимптотической устойчивости некоторых экономичных разностных схем, Ж. вычисл. матем. и матем. физ. **20**, 2 (1980), 350—358.

15 Schmidt, J. W.; Schoenheinz, H., Fehlerschranken zum Differenzenverfahren unter ausschließlicher Benutzung verfügbarer Größen, Arch. Rat. Mech. Analysis 10 (1962), 311—322.

16 Schmidt, J. W., Defektabschätzungen bei Differenzenverfahren, ZAMM **46** (1966), 17—39.

17 Stoyan, G., Towards a general-purpose difference scheme for the linear one-dimensional parabolic equation. In: Nonlinear Analysis. Theory and Applications (ed. R. Kluge), Akademieverlag, Berlin 1981, S. 297—314.

18 Ciarlet, P. G.; Schultz, M. H.; Varga, R. S., Numerical methods of high-order accuracy for nonlinear boundary value problems. I: One dimensional problems, Numer. Math. 9 (1967), 394—430.

19 Perrin, F. M.; Price, H. S.; Varga, R. S., On higher-order numerical methods for nonlinear two-point boundary value problems, Numer. Math. **13** (1969), 180—198.

20 Schultz, M. H.; Varga, R. S., L-splines, Numer. Math. **10** (1967), 345—369.

21 Rose, M. E., Finite difference schemes for differential equations, Math. Comput. 18 (1964), 175—195.

22 Hulme, B. L., Interpolation by Ritz approximation, J. Math. Mech. 18 (1968), 337—342.

23 de Groen, P. P. N.; Hemker, P. W., Error bounds for exponentially fitted Galerkin methods applied to stiff two-point boundary value problems. In: [3], S. 217—249.

24 Ciarlet, P. G.; Schultz, M. H.; Varga, R. S., Numerical methods of high-order accuracy for nonlinear boundary value problems. V: Monotone operator theory, Numer. Math. **13** (1969), 51—77.

25 Bellman, R., On the non-negativity of Green's functions, Boll. d'Unione Mate., Ser. III **12** (1957), 411—413.

26 Bohl, E., Monotonie: Lösbarkeit und Numerik bei Operatorgleichungen, Springer, Berlin 1974.

27 Brandt, A., Multi-level adaptive solutions to boundary-value problems, Math. Comput. **31** (1977), 333—390.

28 Hackbusch, W., On the fast solution of nonlinear elliptic problems, Numer. Math. **32** (1979), 83—95.
29 Keller, H. B.; Cohen, D. S., Some positone problems suggested by nonlinear heat generation, J. Math. Mech. **16** (1967), 1361—1376.
30 Красносельский, М. А.; Вайникко, Г. М.; Забрейко, П. П.; Рутицкий, Я. Б.; Стеценко, В. Я., Приближенное решение операторных уравнений, Наука, Москва 1969.
31 Cohen, D. S.; Laetsch, T. W., Nonlinear boundary value problems suggested by chemical reactor theory, J. Diff. Eqs. 7 (1970), 217—226.
32 Bielecki, A., Une remarque sur la méthode de Banach-Cacciopoli-Tikhonov dans la théorie des équations différentielles ordinaires, Bull. Acad. Polon. Sci., Ser. III **4**, **5** (1956), 261—264.
33 Калиткин, Н. Н., Численные методы, Наука, Москва 1978.

Eingereicht am 26. 10. 1981, endgültige Fassung am 6. 8. 1982

Anschrift: Dr. Gisbert Stoyan, Institut für Mathematik der AdW der DDR, Mohrenstraße 39, DDR-1080 Berlin

KLEINE MITTEILUNGEN

ZAMM 64, 193 – 194 (1984)

V. Komkov

Sensitivity Analysis for Beams and Plates (The Elementary Solution Approach)

In this note we shall first establish reasons for studying sensitivity of dynamic systems subjected to inpulse loads.

We consider the linear equations modelling the dynamic behavior of vibrating beams and thin plates. We develop the basic equations for the sensitivity function indicating the dynamic structural response to changes in design. Specifically, the properties of the sensitivity function are developed in detail for the vibrating beam equation. We conclude that the knowledge of an elementary solution of the beam equation, and of the sensitivity function for the elementary solution can be used to determine the sensitivity function for a class of admissible loads applied to the beam.

0. Introduction

Let us first consider an abstract case. The state of the system is described by a partial differential equation (or a system of equations) of hyperbolic or mixed type

$L(W(x, t; u(x)) = q(x, t, w(x, t; u(x))$

in $(\Omega \times T)$, $W(x, t; u(x)) = \tilde{W}(x, t)$ is an element of an appropriate space of functions (generally a Sobolev space) and u is a design vector $u = u(x)$ obeying certain physically motivated constraints: $\psi_j(u) \leqq 0$, $j = 1, 2, \ldots, t$. $\tilde{W}(x, t)$ obeys boundary conditions

$$B\tilde{W}(x, t) = 0\,, \qquad x \in \partial\Omega\,, \tag{0.1a}$$

and initial conditions $W(x, 0) = W_0(x)$, $x \in \Omega$. $q(x, t; u(x))$ is an element of a topological space determined by physical requirements. Generally, the force $q(x, t; u(x))$ acting on the system is a generalized function; $q \in Q$.

Let $\langle, \rangle_\Omega$ denote the $L_2(\Omega)$ product, or the product of elements of Q acting on Q^*. Looking for solutions in an appropriate Sobolev space we consider a weak form of the equation (0.1)

$$\langle L(W(x, t; u(x)), \lambda\rangle - \langle q(x, t; u(x)), \lambda\rangle = 0\,, \tag{0.1*}$$

where λ is a vector in the space dual to Q, i.e. $\lambda \in Q^*$. (if $Q = H_0^{-s}(\Omega)$ then $\lambda \in H_0^{+s}(\Omega)$). The vector λ may be required to obey (in the weak sense) the adjoint equation $L^*\lambda = 0$, with boundary conditions and finite conditions determined by examining the specific properties of the system (0.1), (0.1a). The specific case which we shall examine is that of the vibrating beam. Let us formally vary the design. Under arbitrary changes in design we preserve the basic equation of motion.

This is done formally by writing

$$\frac{\partial}{\partial u}\{\langle L(u)\, W(u), \lambda\rangle - \langle \hat{q}(u), \lambda\rangle\} = 0\,. \tag{0.1}$$

Symbolically this is expressed by the relation

$$\langle L(u)\, W_u, \lambda\rangle + \langle L_u W, \lambda\rangle - \langle \hat{q}_u, \lambda\rangle = 0\,, \tag{0.2}$$

where W is an element of $H_0^{2,1}$ in the case of a vibrating beam. Here the operator L_u is given by

$$L_u = \frac{\partial^2}{\partial x^2}\left(E\,\frac{\partial I(u)}{\partial u}\,\frac{\partial^2\cdot}{\partial x^2}\right) + \varrho\,\frac{\partial A}{\partial u}\,\frac{\partial^2\cdot}{\partial t^2} \tag{0.3a}$$

for the vibrating beam, and

$$(L_u W) = \left[3h^2\,\frac{\partial h}{\partial u}\,(W_{xx} + \nu W_{yy})\right]_{xx} + \left[3h^2\,\frac{\partial h}{\partial u}\,(W_{yy} + \nu W_{xx})\right]_{yy} + 2(1-\nu)\left[3h^2\,\frac{\partial h}{\partial u}\,W_{xy}\right]_{xy} + \varrho\,\frac{\partial A}{\partial u}\,W_{tt} \tag{0.3b}$$

for the vibrating thin plate.

In general, the operator L_u is constructed by formally differentiating the coefficients of the operator L with respect to the design function u. We shall discuss specifically the beam operator. However, nearly all results generalize to arbitrary hyperbolic operators.

1. Some properties of the sensitivity function

Let us consider the beam equation of motion of the form

$$(EI(u(x), x)\, W_{xx})_{xx} + A(u(x))\, \varrho(x)\, W_{tt} + M(u(x))\, W_t + KW = q(x, t) \tag{1.1}$$

with constraints of the form

$$\psi_j = \int_0^T \int_{-l}^{+l} f_j(W, u)\, dx\, dt \leqq 0\,, \qquad j = 1, 2, \ldots, r\,. \tag{1.1a}$$

We may wish to minimize a functional of the form

$$\psi_0 = \int_0^t \int_{-l}^{+l} f_0(W, u)\, dx\,. \tag{1.2}$$

Here we assume that f_j, $j = 0, 1, \ldots, r$, are differentiable functions of the variables W and u.

The initial conditions assigned to W are $W(x, 0) = \zeta_1(x)$, $W_t(x, 0) = 0$. The beam is simply supported, that is

$$\begin{cases} W(-l, t) = W(l, t) \equiv 0\,, \\ EIW_{xx}(-l, t) = EIW_{xx}(+l, t) \equiv 0 \quad \text{for all values of } t\,. \end{cases}$$

We introduce the displacement sensitivity function S_u, representing the sensitivity of the state function with respect to changes in design. This function is crucial in sensitivity analysis, but difficult to determine without solving the basic equation of state. We assume that the function S_u obeys the following functional relation

$$\langle S_u(x, t, u(x)), u(x), \Delta u(x)\rangle_\Omega = W(x, t, U(x) + \Delta u(x)) - W(x, t, u(x)) + r(\Delta u^2, \ldots)\,, \tag{1.3}$$

and that

$$\lim_{\|\Delta u\| \to 0} \frac{|r((\Delta u)^2 \ldots)|}{\|\Delta u\|} = 0\,. \tag{1.3a}$$

In other words, we assume Fréchet differentiability of the displacement function W with respect to the distributed parameter vector function $u(x)$. If u is a single (constant) parameter then $S_u = \dfrac{\partial W}{\partial u}$.

Let us copy the usual procedure of sensitivity analysis. We multiply both sides of the state equation (1.1) by some function $\lambda \in H_0^{2,1}(\Omega)$ and integrate over Ω, obtaining

$$\langle (EI(u(x), x)\, W_{xx})_{xx}, \lambda\rangle_\Omega + \langle A(u(x), x), \varrho(x)\, W_{tt}, \lambda\rangle_\Omega + \langle \mu(u(x))\, W_t, \lambda\rangle_\Omega + K \cdot \langle W, \lambda\rangle - \langle q(x, t), \lambda\rangle = 0\,.$$

The parameter variation problem may be regarded as the problem of Bolza, that is of varying the functional ψ_0 subject to constraints (1.1a). We also regard the state equation (1.1) as an added constraint condition. Subject to some smoothness of f_j, we can restate the variational problem in the form given below. Analyze the variation of the functional:

$$J = \psi_0 \sum_{j=1}^{r} K_j \psi_j + \langle EI(u(x), x)\, W_{xx}, \lambda_{xx}\rangle_\Omega - \langle A(u(x), x)\, \varrho(x)\, W_t, \lambda_t\rangle_\Omega - \langle \mu(u(x))\, W, \lambda_t\rangle_\Omega + K \cdot \langle W, \lambda\rangle_\Omega - \langle q(x, t), \lambda\rangle_\Omega + \alpha\langle (W - \zeta_1(x)), \lambda\rangle_{\partial\Omega_1} + \beta\langle \dot{W}, \lambda\rangle_{\partial\Omega_1} + \gamma\langle W, \lambda\rangle_{\partial\Omega_2} + \delta\langle M_w, \lambda\rangle_{\partial\Omega_2}\,. \tag{1.4}$$

We have constructed the functional J by utilizing integration by parts and assuming that the following boundary terms vanish:

$\langle (EI(u(x), x)\, W_{xx})_x \cdot \lambda\rangle_{\partial\Omega_2} = 0\,,$

$\langle A(u(x), x)\, \varrho(x)\, W_t \cdot \lambda\rangle_{\partial\Omega \times T} = 0\,.$

These boundary terms vanish if we assign

$\lambda|_{\partial\Omega_2} \equiv 0\,, \qquad \lambda(x, T) \equiv 0\,.$

We carry out the rules of Fréchet differentiation for the functional J:

$$\left.\frac{\partial J}{\partial \lambda}\right|_\Omega = (EI(u, (x), x)\, W_{xx})_{xx} + A(u; x)\, \varrho W_{tt} + \mu(u) W_t + Kw - q(x, t)\,, \tag{1.5a}$$

$$\left.\frac{\partial J}{\partial \lambda}\right|_{\partial\Omega_1} \doteq \alpha(W(x,0) - \zeta_1(x)) + \beta(\dot{W}(x,0)) , \tag{1.5b}$$

$$\left.\frac{\partial J}{\partial \lambda}\right|_{\partial\Omega_2} = [\gamma W + \delta M_w]_{\partial\Omega_2} , \tag{1.5c}$$

where α, β, γ, δ are Lagrangian multipliers.

Setting $\partial J/\partial \lambda = 0$ we recover our equation of state and the boundary and initial conditions:

$$W(x,0) = \zeta_1(x), \qquad \dot{W}(x,0) = 0 ,$$

$$W\,|_{\partial\Omega_2=(\pm l\times T)} = 0 , \qquad M_W = EIW_{xx}|_{\partial\Omega_2=(\pm l\times T)} = 0 .$$

Differentiation of J with respect to W is the key to sensitivity analysis.

$$\frac{\partial J}{\partial W} = \frac{\partial f_0}{\partial W} - \sum_{j=1}^{r} K_j \frac{\partial f_j}{\partial W} + (EI(u;x)\,\lambda_{xx})_{xx} +$$
$$+ \varrho A(u;x)\,\lambda_{tt} - \mu(u(x))\,\lambda_t + K\lambda = 0 \quad \text{in } \Omega , \tag{1.6}$$

$\lambda(x,t) = 0$ on the $\partial\Omega_1$ part of the boundary,

$\lambda \equiv 0$ on $\partial\Omega_2$,

$\lambda \equiv 0_{\partial\Omega_2 \cup \partial\Omega_2 = \{\pm l\times[0,\tau]\}}$.

The Fréchet chain rule of differentiation may be stated as follows: J_u is a vector obeying the formula $J_u = J_W \dfrac{\partial W}{\partial u} + \dfrac{\partial J}{\partial u}$,

where

$$J(u + \Delta u) - J(u) = \langle J_u, \Delta u\rangle + r(\Delta u^2)$$

f $r(\Delta u^2)$ has the property (1.3a). Therefore

$$J_u = \left[\left(\frac{\partial f_0}{\partial W} - \sum_{j=1}^{r} K_j \frac{\partial f_j}{\partial W}\right)\right] S_u + [(EI\lambda_{xx})_{xx} + A\varrho\lambda_{tt} -$$
$$- \mu\lambda_t + K\lambda)]\, S_u . \tag{1.7}$$

We note that $J_u \equiv 0$ on $\partial\Omega$ because of conditions assigned to λ and W.

The situation becomes very simple if u is a lumped parameter and $S_u = \partial W/\partial u$. We can directly differentiate the functional J with respect to u obtaining

$$\left.\frac{\partial J}{\partial u}\right|_{\Omega} = \left\langle E\frac{\partial I}{\partial u} W_{xx}, \lambda_{xx}\right\rangle + \left\langle EI(u)\left(\frac{\partial W}{\partial u}\right)_{xx}, \lambda_{xx}\right\rangle$$
$$- \left\langle\frac{\partial A}{\partial u}\varrho\cdot W_t, \lambda_t\right\rangle - \left\langle \varrho A\left(\frac{\partial W}{\partial u}\right)_t, \lambda_t\right\rangle -$$
$$- \left\langle\left(\frac{\partial\mu}{\partial u}\right) W, \lambda_t\right\rangle + \left\langle K\left(\frac{\partial W}{\partial u}\right), \lambda\right\rangle - K_j\frac{\partial fj}{\partial W}\cdot\frac{\partial W}{\partial u} \text{ in } \Omega. \tag{1.8}$$

Hence, S_u obeys the differential equation

$$E(I(u)\,(S_u)_{xx})_{xx} + \varrho A(u)\,(S_u)_{tt} + \mu(S_u)_t + KS_u +$$
$$- \sum_{j=1}^{r} K_j \frac{\partial f_j}{\partial W} S_u = \eta(W) , \tag{1.9}$$

where

$$\eta(W) = E\left(\frac{\partial I(u)}{\partial u} W_{xx}\right)_{xx} + \varrho\left(\frac{\partial A}{\partial u}\right) W_{tt} + \frac{\partial\mu}{\partial u} W .$$

Let us consider the design of a beam and assume that I and A are monotone functions of u and $\partial I/\partial u > 0$, $\partial A/\partial u > 0$ and that $\partial\mu/\partial u \geqq 0$. These are physically reasonable assumptions if u denotes some basic dimension of the cross-section. Therefore, S_u obeys a beam equation with elastic support term that contains terms dependent on the constraint functions and with an inhomogeneous term depending on the solution of another beam equation.

In the case when $K = 0$, and $\mu = 0$, we have

$$J_u = \left[\frac{\partial f_0}{\partial W} - \sum_{j=1}^{r} K_j \frac{\partial f_j}{\partial W}\right] S_u + [(EI\lambda_{xx})_{xx} + A\varrho\lambda_{tt}]\, S_u , \tag{1.10}$$

and S_u obeys the differential equation

$$-[(EI(u)\,S_{u_{xx}})_{xx} + \varrho A(u)\,S_{u_{tt}}] = L_u W = \left(E\frac{\partial I}{\partial u} W_{xx}\right)_{xx} +$$
$$+ \varrho\frac{\partial A}{\partial u} W_{tt} , \tag{1.11}$$

S_u satisfies the boundary conditions:

$$S_u(x=0) = S_u(x=l) = 0 , \tag{1.11a}$$

$$EIS_{u_{xx}}(x=0) = EIS_{u_{xx}}(x=l) = 0 , \tag{1.11b}$$

for a simply supported beam, and the initial condition

$$S_u(x, (t=0)) = 0 , \qquad (S_u)_t\,(x, (t=0)) = 0 . \tag{1.11c}$$

In a statically determinate case the bending moment is independent of the design, that is: EIW_{xx} is not a function of the design vector of u. Consequently

$$\frac{\partial}{\partial u}(EIW_{xx}) = 0 , \quad \text{and} \quad E\frac{\partial I}{\partial u} W_{xx} = -EI(S_u)_{xx} . \tag{1.12}$$

Substituting this expression into (1.11) we have

$$\varrho A(u)\,(S_u)_{tt} = 2\left(E\frac{\partial I}{\partial u} W_{xx}\right)_{xx} + \varrho\frac{\partial A}{\partial u} W_{tt} .$$

Therefore

$$S_u = t * \left\{(\varrho A)^{-1}\left[2E\left(\frac{\partial I}{\partial u} W_{xx}\right)_{xx} + \varrho\frac{\partial A}{\partial u} W_{tt}\right]\right\} \tag{1.13}$$

where $*$ denotes the convolution operation

$$f * g = \int_0^t f(t-\tau)\, g(\tau)\, d\tau .$$

The formulas (1.11)–(1.13) indicate that one can determine the properties of the sensitivity function S_u as well as the properties for certain classes of solutions of the inhomogeneous beam equation, if only one knows the elementary solution of the beam equation, $e(x,t)$ satisfying boundary conditions (1.11a), (1.11b) and the equation

$$(EI(u)\,e_{xx})_{xx} + \varrho A(u)\,e_{tt} = \delta(x,t) \tag{1.14}$$

for suitable functions I and A. Here $\delta(x,t)$ denotes the Dirac delta function, that is $\delta(x,t)$ is a distribution over $C^{\infty}_{[0,l]\times[0,\tau]}$ functions of x and t, such that

$$\int_{-\varepsilon}^{+\varepsilon}\int_{-\varepsilon}^{+\varepsilon} \delta(x,t)\, dx\, dt = 1 , \quad \text{for any } \varepsilon > 0 ,$$

and

$$\int\int_{[0,l]\times[0,\tau]} \delta(x,t)\, f(x,t) = f(0,0) .$$

The function S_u can be computed for a given admissible state function $W(x,t)$ by first computing the right hand side of (1.11) and convoluting it with $e(x,t)$. Almost identical conclusions can be reached for plate theory, except that the simplifications made for statically determined cases do not apply to plates with usual types of boundary support.

The solution $e(x,t)$ of the equation (1.14) is called the elementary solution to the beam equation. Usually the convolution algebra is not introduced in a rigorous manner in engineering monographs, the convolution kernels being introduced as functions of multiple variables, such as $e = e(x,\xi,t,\tau)$, etc. If we ignore the time effects, then in readily understood civil engineering terms e is the influence function $e(x,\xi)$, describing the effects at coordinate x due to a unit point load (Dirac delta function) applied at the coordinate ξ. If we ignore the spatial effects and consider only a unit (time) impulse $\delta(t)$, then $e(t,\tau)$ assumes a familiar meaning to the electrical engineers studying impedence networks.

Note: A completely different approach to the problem of sensitivity for beams and plates is pursued in current papers by E. J. Haug and V. Komkov. Their results will be summarized in [1].

References

1 Haug, E. J.; Choi, K.; Komkov, V., Sensitivity Analysis for Structural and Mechanical Systems, monograph to be published by Academic Press, New York 1983.

Received July 7, 1982

Author's address: Professor Dr. Vadim Komkov, Department of Mathematics, Winthrop College, Rock Hill, S. C. 29733 U.S.A.

ZAMM 64, 195 – 196 (1984)

P. Filip / V. Kolář / A. G. Curev

The Plane Jet as a Limit Case of the Radial Jet

1. Introduction

The incompressible plane jet ranks both theoretically and experimentally among the most investigated problems in the sphere of submerged jets. This problem is described by the equation of motion (for simplicity of notation we do not introduce a different notation for the time-averaged magnitudes in a turbulent case)

$$u\frac{\partial u}{\partial x} + v\frac{\partial u}{\partial y} = \nu\frac{\partial^2 u}{\partial y^2} \quad \text{(for a laminar case)} \tag{1a}$$

or

$$u\frac{\partial u}{\partial x} + v\frac{\partial u}{\partial y} = \frac{1}{\varrho}\frac{\partial \tau}{\partial y} \quad \text{(for a turbulent case)}, \tag{1b}$$

by the continuity equation

$$\frac{\partial u}{\partial x} + \frac{\partial v}{\partial y} = 0, \tag{2}$$

and by the conditions

$$\lim_{y\to\pm\infty} u(x, y) = 0, \qquad \lim_{y\to\pm\infty} \frac{\partial u}{\partial y}(x, y) = 0, \tag{3}$$

$$\frac{\partial u}{\partial y}(x, 0) = 0, \tag{4}$$

$$v(x, 0) = 0 \tag{5}$$

where $x \in (0, +\infty)$.

Let us denote u_{PL} and u_{PT} the solution of a laminar case (1a), (2)–(5) and of a turbulent case (1b), (2)–(5) of the plane jet, respectively. According to Schlichting [6] we have

$$u_{PL}(x, y) = \frac{1}{2}\left(\frac{3M_P^2}{\varrho^2\nu}\right)^{1/3} x^{-1/3}(1 - \tanh^2 \eta_{PL}) \tag{6a}$$

where

$$\eta_{PL}(x, y) = \frac{1}{2}\left(\frac{M_P}{3\varrho\nu^2}\right)^{1/3} x^{-2/3}y; \tag{7a}$$

$$u_{PT}(x, y) = \frac{1}{2}\left(\frac{3M_P}{2k\varrho}\right)^{1/2} x^{-1/2}(1 - \tanh^2 \eta_{PT}) \tag{6b}$$

where

$$\eta_{PT}(x, y) = \frac{1}{4k}\cdot\frac{y}{x}. \tag{7b}$$

The constant M_P represents the integral invariant of the conservation of the momentum flux

$$M_P = \int_0^\infty \varrho u_{PL}^2 \, dy, \quad \text{and} \quad M_P = \int_0^\infty \varrho u_{PT}^2 \, dy, \tag{8a, b}$$

respectively, and results in the same way for both regimes of a flow from (1)–(5). k is an experimental constant.

The incompressible radial jet is governed by the equation of motion (1a) and (1b), respectively (where x denotes – as in the plane problem – a distance from a point-source in the momentum flux direction and y represents a transverse coordinate), by the continuity equation

$$\frac{\partial}{\partial x}(xu) + \frac{\partial}{\partial y}(xv) = 0, \tag{9}$$

and by the conditions (3)–(5). The basic solution of the radial jet is shown in Riley [5], Schwarz [7] in the form

$$u_{RL}(x, y) = \frac{1}{2}\left(\frac{9M_R^2}{\varrho^2\nu}\right)^{1/3} x^{-1}(1 - \tanh^2 \eta_{RL}) \tag{10a}$$

where

$$\eta_{RL}(x\ y) = \frac{1}{2}\left(\frac{3M_R}{\varrho\nu^2}\right)^{1/3} \frac{y}{x}; \tag{11a}$$

$$u_{RT}(x, y) = \frac{1}{2}\left(\frac{3M_R}{k\varrho}\right)^{1/2} x^{-1}(1 - \tanh^2 \eta_{RT}) \tag{10b}$$

where

$$\eta_{RT}(x, y) = \frac{1}{2k}\cdot\frac{y}{x}. \tag{11b}$$

The constant M_R denotes the integral invariant of the conservation of the radial momentum flux

$$M_R = \int_0^\infty \varrho x u_{RL}^2 \, dy, \quad \text{and} \quad M_R = \int_0^\infty \varrho x u_{RT}^2 \, dy, \tag{12a, b}$$

respectively, following in the same way for both regimes of a flow from (1), (3)–(5), (9). k is an experimental constant.

2. The Limiting Process

The notion of a virtual origin of a submerged jet is introduced by Abramovich [1] and Hinze [2]. It is connected with a higher accuracy of a similarity solution for the description of a jet ejected from a source of finite dimensions by the introduction of a new experimental constant, which represents the location of an adequate point-source usually called the virtual origin of a submerged jet.

In the case of the radial jet the determination of the virtual origin does not appear as a mere translation as in the simple classical problems (e.g. the plane and axisymmetric jets) but it bears in this case a more complicated functional dependence determined by the regime of a flow. According to Riley [5], Schwarz [7] we have for a laminar case the generalized similarity solution

$$u_{RL}(x, y) = \frac{1}{2}\left(\frac{9M_R^2}{\varrho^2\nu}\right)^{1/3} (x^3 - l^3)^{-1/3}(1 - \tanh^2 \eta_{RL}) \tag{13a}$$

where

$$\eta_{RL}(x, y) = \frac{1}{2}\left(\frac{3M_R}{\varrho\nu^2}\right)^{1/3} x(x^3 - l^3)^{-2/3}y, \tag{14a}$$

and according to Kolář, Filip, Curev [3] for a turbulent case

$$u_{RT}(x, y) = \frac{1}{2}\left(\frac{3M_R}{k\varrho}\right)^{1/2} (x^2 - l^2)^{-1/2}(1 - \tanh^2 \eta_{RT}) \tag{13b}$$

where

$$\eta_{RT}(x, y) = \frac{1}{2k}\cdot\frac{x}{x^2 - l^2}\cdot y \tag{14b}$$

where x denotes a distance from the axis of a jet symmetry, y represents again a transverse coordinate.

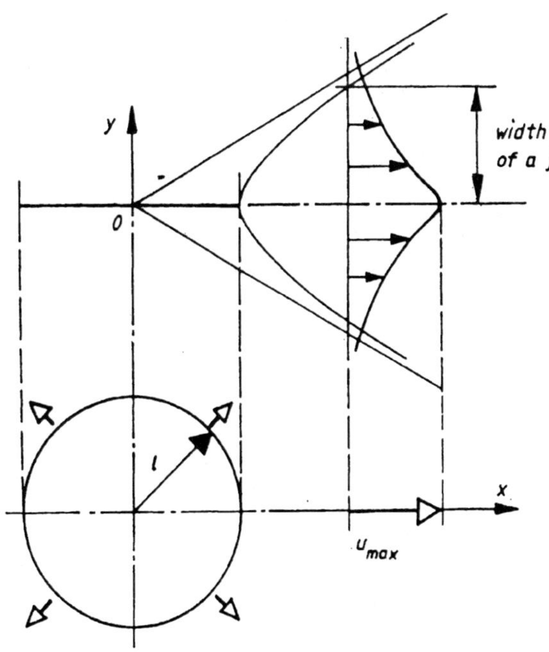

Fig. 1. The circumference-point-source

Riley [5] introduces a new constant l as a characteristic length associated with the boundary layer development, Schwarz [7] introduces l as a constant giving more accuracy to the basic solution near the origin. With regard to the conditions and the notion of a point-source we may in a unique way characterize l as the radius of a circle where the point-sources are continuously located on its circumference. By this interpretation we obtain the notion of a circumference-point-source of a radius l, see Fig. 1. It means that the physical interpretation of a characteristic length l of a generalized similarity solution represents for the given type of a jet (the radial jet) the adequate point-source (the circumference-point-source). This consideration reflects the primary interpretation of the constant l. Obviously a circumference-point-source reduces to a point one for the limit case $l \to 0$. This fact mathematically means the reduction of the relations (13), (14) to the relations (10), (11).

In the following we show the passage of (13), (14) to the limit case $l \to +\infty$ which does not seem to be so apparent as in the case $l \to 0$.

Let us write the relation (12a) in the form

$$\frac{\partial}{\partial x} \int_0^\infty \varrho x u_{RL}^2 \, dy = 0 \tag{15}$$

and after rearranging

$$\frac{\partial}{\partial x} \int_0^\infty \varrho u_{RL}^2 \, dy + \frac{M_R}{x^2} = 0 \,. \tag{16}$$

With respect to a physical reasoning we may suppose $x > l$. If we fix $x - l$, we get from the above expression (16) by a limiting procedure for $l \to \infty$

$$\int_0^\infty \varrho (\lim_{l\to\infty} u_{RL})^2 \, dy = \overline{M}_P \tag{17}$$

where $\overline{M}_P$ is a positive constant.

Let us show a limiting procedure in the relation (14a) where $\bar{x}$ denotes a finite (fixed) difference $x - l$.

$$\begin{aligned}\lim_{\to\infty} \eta_{RL}(x, y) &= \frac{1}{2}\left(\frac{3}{\varrho\nu^2}\right)^{1/3} \cdot \left[\lim_{l\to\infty}\left(x \int_0^\infty \varrho u_{RL}^2 \, dy\right)\right]^{1/3} \cdot x \cdot \\ &\quad \cdot \left[\lim_{l\to\infty} (x - l)(x^2 + lx + l^2)\right]^{-2/3} \cdot y \\ &= \frac{1}{2}\left(\frac{3}{\varrho\nu^2}\right)^{1/3} \cdot \left[\int_0^\infty \varrho \left(\lim_{l\to\infty} u_{RL}\right)^2 dy\right]^{1/3} \cdot \bar{x}^{-2/3} \cdot \\ &\quad \left[\lim_{l\to\infty}\left(1 + \frac{l}{x} + \left(\frac{l}{x}\right)^2\right)\right]^{-2/3} \cdot y = \\ &= \frac{1}{2}\left(\frac{3\overline{M}_P}{\varrho\nu^2}\right)^{1/3} \cdot \bar{x}^{-2/3} \cdot 3^{-2/3} \cdot y \,,\end{aligned}$$

$$\lim_{l\to\infty} \eta_{RL}(x, y) = \frac{1}{2}\left(\frac{\overline{M}_P}{3\varrho\nu^2}\right)^{1/3} \cdot \bar{x}^{-2/3} \cdot y \,. \tag{18}$$

In a similar way we obtain from (13a)

$$\begin{aligned}\lim_{l\to\infty} u_{RL}(x, y) &= \frac{1}{2}\left(\frac{9}{\varrho^2\nu}\right)^{1/3} \cdot \left[\lim_{l\to\infty} x \int_0^\infty \varrho u_{RL}^2 \, dy\right]^{2/3} \\ &\quad \cdot \left[\lim_{l\to\infty} (x - l)(x^2 + lx + l^2)\right]^{-1/3} \cdot \\ &\quad \cdot \left[1 - \tanh^2 \left(\lim_{l\to\infty} \eta_{RL}\right)\right] = \\ &= \frac{1}{2}\left(\frac{9}{\varrho^2\nu}\right)^{1/3} \cdot \overline{M}_P^{2/3} \cdot 3^{-1/3} \cdot \bar{x}^{-1/3} \cdot \\ &\quad \cdot \left[1 - \tanh^2\left(\frac{1}{2}\left(\frac{\overline{M}_P}{3\varrho\nu^2}\right)^{1/3} \cdot \bar{x}^{-2/3} \cdot y\right)\right],\end{aligned}$$

$$\begin{aligned}\lim_{l\to\infty} u_{RL}(x, y) &= \frac{1}{2}\left(\frac{3\overline{M}_P^2}{\varrho^2\nu}\right)^{1/3} \cdot \bar{x}^{-1/3} \cdot \\ &\quad \left[1 - \tanh^2\left(\frac{1}{2}\left(\frac{\overline{M}_P}{3\varrho\nu^2}\right)^{1/3} \cdot \bar{x}^{-2/3} \cdot y\right)\right].\end{aligned} \tag{19}$$

From here it is evident by comparison (18), (19) with (6a), (7a)

$$\lim_{l\to\infty} u_{RL}(x, y) = u_{PL}(\bar{x}, y) \,. \tag{20}$$

As the difference $\bar{x} = x - l$ may be fixed arbitrarily, i.e. $\bar{x} \in (0, +\infty)$, it follows that the relation (20) is valid for all positive $\bar{x}$. Here $\bar{x}$ represents a distance from a source in the momentum flux direction (denoted by x in the Introduction) for a plane problem, a constant $\overline{M}_P = M_P$.

We obtain the relations (6b), (7b) for the turbulent case from (13b), (14b) by the same limiting procedure as in the laminar case and we get

$$\lim_{l\to\infty} u_{RT}(x, y) = u_{PT}(\bar{x}, y) \,. \tag{21}$$

For simplicity the notation $x = x - l$ was introduced. It is evident that the limiting process in (18), (19) with the notation $\bar{x} - L = x - l$ holds its validity and in such a way we may proceed to a generalized similarity solution of a plane jet (where $\bar{x} - L$ denotes a distance from a source in the momentum flux direction).

3. Conclusion

From the above it follows that the description of a velocity field of the plane jet is obtainable by a limiting process of a generalized similarity description of a radial jet velocity field. It means that the plane jet is a limit case of the radial jet. The limiting process is based on a proper interpretation of a constant l in a generalized similarity solution of a radial jet, i.e. on the concept of a circumference-point-source of a radius $l \in [0, +\infty)$ and follows from the expansion of the validity of this concept to the closed interval $l \in [0, +\infty]$.

References

1 Abramovich, G. N., The Theory of Turbulent Jets, M.I.T. Press, Massachusetts, 1963.
2 Hinze, J. O., Turbulence, McGraw-Hill, New York/Toronto/London, 1959.
3 Kolář, V.; Filip, P.; Curev, A. G., Comments on the radial turbulent jet, Acta Technica ČSAV 27 (1982), 563–567.
4 Rajaratnam, N., Turbulent Jets, Developments in Water Science No. 5, Elsevier, Amsterdam/Oxford/New York, 1976.
5 Riley, N., Asymptotic Expansions in Radial Jets, J. Math. Phys. 41 (1962), 132–146.
6 Schlichting, H., Boundary Layer Theory, McGraw-Hill, New York, 1968.
7 Schwarz, W. H., The radial free jet, Chem. Eng. Sci. 18 (1963), 779–786.

Received 1983 Jan. 13

Address: P. Filip, CSc; Ing. V. Kolář; Ing. A. G. Curev, DrSc, Institute of Hydrodynamics, Czechoslovak Academy of Sciences, Podbabská 13, 166 12 Prague 6, Czechoslovakia

ZAMM 64, 196–198 (1984)

N. Phan-Thien / D. K. Gartling

A Rayleigh Flow of a Plastic Fluid for an Arbitrarily Prescribed Boundary Shear Stress

Introduction

Many fluids used in the food and oil drilling industries behave as if they possess a yield shear stress. That is, below this yield shear stress they do not deform, at least in the laboratory time scale; and above this yield shear stress, they flow like Newtonian fluids, i.e. ones with constant viscosities. Because of the nonzero yield stress, the flow of any plastic fluid involves two regions: a "no-flow" region where the "shear stress" is less than the yield shear stress and the "flow" region where the "shear stress" is greater than the yield shear stress. Both regions are separated by a yield surface whose position is generally time-dependent and must be found as part of the solution. Because of the presence of the moving yield surface, any flow problem of a plastic fluid is nonlinear and may be best solved numerically. One such finite-element program designed to solve incompressible, viscous, plastic flow problems with inertia has been developed by the second author [1] and need not be considered here. Instead, this note is concerned with the flow of a plastic fluid which occupies the half space $y \geq 0$ and which is driven by an arbitrarily time-dependent prescribed shear stress at the boundary $y = 0$. The case where the applied boundary shear stress is a constant has been shown to possess a similarity solution [2]. The present work is an extension of [2] using the method developed by Tao [3] in solving moving boundary problems in half space. These half-space problems, apart from being of interest themselves, form convenient benchmark problems for numerical codes.

Analysis

We consider the flow of a plastic fluid above the plane $y = 0$ which is generated at time $t > 0$ by imparting a shear stress $T(t)$ on the boundary plane $y = 0$ (y is the direction normal to the boundary). As in TAO [3], we assume that $T(\cdot)$ is analytic in its argument and thus can be expanded as a power series in the time. This is a unidirectional flow where the velocity field is given by $(u(y, t), ,0, 0)$. One of the simplest constitute relations for the fluid in this flow is

$$\tau(y,t) = \begin{cases} \eta_1 \dfrac{\partial u_1}{\partial y}(y,t), & 0 \leqq y \leqq s(t), \\[2ex] \eta_2 \dfrac{\partial u_2}{\partial y}(y,t), & s(t) \leqq y < \infty, \end{cases} \tag{1}$$

where $\tau(y, t)$ is the shear stress at y and time t; η_i are material constants of dimensions of viscosity and u_1, u_2, respectively, are the x-component velocities in the "flow" and in the "no-flow" regions. Note that in the flow region, $0 \leqq y \leqq s(t)$, we have $|\tau(y, t)| \geqq \tau_y$, where τ_y is the material yield shear stress. At the yield surface $y = s(t)$, the shear stress is given by τ_y:

$$\eta_1 \frac{\partial u_1}{\partial y}(s(t), t) = \eta_2 \frac{\partial u_2}{\partial y}(s(t), t) = \tau_y \,. \tag{2}$$

We also rule out the presence of a vortex sheet at $y = s(t)$. Thus,

$$u_1(s(t), t) = u_2(s(t), t) \,. \tag{3}$$

For finite values of η_1 and η_2, there will be some deformation in the "no-flow" region $s(t) \leqq y$ where $|\tau(y, t)| \leqq \tau_y$. This deformation will be small for a real plastic fluid for which η_2/η_1 is large. For an ideal BINGHAM fluid [4], $\eta_2/\eta_1 = \infty$. (In the finite element code of GARTLING [1], the value of η_2/η_1 can be prescribed arbitrarily. The constitutive relation for the fluid, however, is the full three-dimensional generalization of (1), in the OLDROYD's sense [5].)

Denoting the shear stresses in the flow and no-flow regions by τ_1 and τ_2, respectively, then both τ_i are governed by the heat diffusion equation

$$\frac{\partial \tau_i}{\partial t} = \nu_i \frac{\partial^2 \tau_i}{\partial y^2} \,, \tag{4}$$

where $\nu_i = \eta_i/\varrho$ are the kinematic viscosities, and ϱ is the common fluid density in both regions. In (4), we have neglected body forces. The boundary conditions for (4) are

$$\tau_1(0, t) = T(t) \,, \qquad \tau_2(y, 0) = 0 = \tau_2(\infty, t) \tag{5}$$

while the interfacial boundary conditions are given by (2) and (3). This completes the formulation of the flow problem. Note that the present problem differs from the classical Stefan problems considered by TAO [3] in the interfacial boundary condition (3). In the latter problems, (3) is replaced by a linear combination of the derivatives $\partial\tau_i/\partial y(s(t), t))$ and $\tau_i(s(t)\, t)$ and ds/dt.

The solution procedure is essentially that of TAO [3] and depends on the following functions

$$\left.\begin{aligned} E_n(\xi) &= \tfrac{1}{2}\,(I^n \operatorname{erfc}(-\xi) + I^n \operatorname{erfc} \xi) \,, \\ F_n(\xi) &= \tfrac{1}{2}\,(I^n \operatorname{erfc}(-\xi) - I^n \operatorname{erfc} \xi) \,, \\ G_n(\xi) &= \tfrac{1}{2}\,(I^n \operatorname{erfc}(-\xi) + (-1)^n I^n \operatorname{erfc} \xi) \,, \end{aligned}\right\} \tag{6}$$

where $\xi_i = y/\sqrt{4\nu_i t}$ is a similarity variable, erfc (ξ) is the complementary error function

$$\operatorname{erfc}(\xi) = 1 - \operatorname{erf}(\xi) = \frac{2}{\sqrt{\pi}} \int_\xi^\infty \exp(-\zeta^2)\, d\zeta \,,$$

and I^n erfc (ξ) are repeated integrals of the complementary error function, defined recurrently by

$$I^n \operatorname{erfc} \xi = \int_\xi^\infty I^{n-1} \operatorname{erfc}(\zeta)\, d\zeta \,, \qquad I^0 \operatorname{erfc} \xi = \operatorname{erfc} \xi \,.$$

The properties of E_n, F_n and G_n have been studied exhaustively in TAO [3]; we list here a few results:

$$I \operatorname{erfc} \xi = \frac{1}{\sqrt{\pi}} e^{-\xi^2} - \xi \operatorname{erfc} \xi \,,$$

$$2n I^n \operatorname{erfc} \xi = I^{n-2} \operatorname{erfc} \xi - 2\xi I^{n-1} \operatorname{erfc} \xi \,,$$

$$2nG_n(\xi) = 2\xi G_{n-1}(\xi) + G_{n-2}(\xi); \qquad G_0(\xi) = 1 \,, \qquad G_1(\xi) = \xi \,,$$

$$d^m G_n(\xi)/d\xi^m = G_{n-m}(\xi) \,,$$

$$dE_n(\xi)/d\xi = F_{n-1}(\xi) \,,$$

$$dF_n(\xi)/d\xi = E_{n-1}(\xi) \,.$$

To generate the solution, we assume the analyticity of $T(\cdot)$ and write

$$T(t) = \tau_y \sum_{n=0}^{\infty} \frac{T_n t^n}{n!} \,, \qquad T_n = \frac{d^n}{dt^n}\left(\frac{T}{\tau_y}\right)\bigg|_{t=0} \,. \tag{7}$$

Since $t^{n/2} I^n$ erfc (ξ_i) is a solution to the heat equation (4), we can represent the solutions by

$$\left.\begin{aligned} \tau_1 &= \tau_y \sum_{n=0}^{\infty} \{T_n (4t)^n G_{2n}(\xi_1) + a_n (4t)^{n/2} F_n(\xi_1) \,, \\ \tau_2 &= \tau_y \sum_{n=0}^{\infty} b_n (4t)^{n/2} I^n \operatorname{erfc}(\xi_2) \,, \end{aligned}\right\} \tag{8}$$

where $\xi_i = y/\sqrt{4\nu_i t}$, a_n and b_n are some constants to be determined.

Because of $G_{2n}(0) = 1/4^n n!$, $F_n(0) = 0$ and erfc $(\infty) = 0$, the solutions (8) satisfy the boundary conditions (5) identically. It remains to satisfy the interfacial boundary conditions (2) and (3). In order to do this, as in TAO [3], we assume that the position of yield surface propagates in time according to

$$s(t) = \sqrt{4\nu_1 t}\, X(t) \,,$$

where

$$X(\mu) = \sum_{n=0}^{\infty} c_n \mu^n \,, \qquad \mu \equiv \sqrt{t} \,. \tag{9}$$

The interfacial boundary conditions (2) lead to

$$\sum_{n=0}^{\infty} \{T_n (2\mu)^n G_{2n}(X(\mu)) + a_n (2\mu)^n F_n(X(\mu))\} = 1 =$$

$$= \sum_{n=0}^{\infty} b_n (2\mu)^n I^n \operatorname{erfc}(\omega X(\mu)) \,, \tag{10}$$

where

$$\omega = \sqrt{\nu_1/\nu_2} = \sqrt{\eta_1/\eta_2} \,.$$

To satisfy the interfacial boundary conditions (3), we note that the velocities are given by, from (1) and (8) using the properties of E_n, F_n and G_n,

$$\begin{aligned} u_1(y,t) &= \frac{\tau_y}{\sqrt{\varrho\eta_1}} \sum_{n=0}^{\infty} \{T_n (2\mu)^{2n+1} G_{2n+1}(\xi_1) + \\ &\quad + a_n (2\mu)^{n+1} E_{n+1}(\xi_1)\} \,, \end{aligned} \tag{11}$$

$$u_2(y,t) = -\frac{\tau_y}{\sqrt{\varrho\eta_2}} \sum_{n=0}^{\infty} b_n (2\mu)^{n+1} I^{n+1} \operatorname{erfc}(\xi_2) \,.$$

Note that both u_i satisfy their respective heat equation. Thus, the interfacial boundary conditions (3) lead to

$$\sum_{n=0}^{\infty} \{T_n (2\mu)^{2n+1} G_{2n+1}(X(\mu)) + a_n (2\mu)^{n+1} E_{n+1}(X(\mu))\} =$$

$$= -\omega \sum_{n=0}^{\infty} b (2\mu)^{n+1} I^{n+1} \operatorname{erfc}(\omega X(\mu)) \,. \tag{12}$$

At this point, it should be noted that the fluid is moving with velocity $U(t)$ at $y = 0$, where

$$U(t) = \frac{\tau_y}{\sqrt{\varrho\eta_1}} \sum_{n=0}^{\infty} a_n (2\mu)^{n+1} E_{n+1}(0) \,, \tag{13}$$

viz. a power series in $\sqrt{t}$, starting with $\sqrt{t}$. The case of $U(t) =$ = constant (Rayleigh's flow) is thus excluded from and cannot be treated by the present solution scheme. One can also show that there is no similarity solution for the case of $U(t) =$ constant.

Since (10) and (12) are identities in $\mu = \sqrt{t}$, they can be expanded in power series in μ and the coefficients a_n, b_n and c_n are determined by matching the power of μ. Equivalently, (10) and (12) can be differentiated with respect to μ and evaluated at $\mu = 0$. Thus, denoting by $D_\mu^N(\cdot) = \frac{\mathrm{d}^N}{\mathrm{d}\mu^N}(\cdot)|_{\mu=0}$, one has [3]

$$D_\mu^N(2\mu)^n\, G_n[\omega X(\mu)] = N!A_n^N(\omega)\,,$$

$$D_\mu^N(2\mu)^n\, I^n \operatorname{erfc}[\omega X(\mu)] = N!B_n^N(\omega)\,,$$

where

$$A_n^N(\omega) = 2^n \sum_{r=0}^{\mu_1} \omega^r G_{n-r}(\omega c_0) \sum_{\alpha_i} \frac{c_1^{\alpha_1} c_2^{\alpha_2} \ldots c_\mu^{\alpha_\mu}}{\alpha_1!\,\alpha_2!\ldots\alpha_\mu!} \tag{14}$$

and

$$B_n^N(\omega) = 2^n \sum_{r=0}^{N-n} (-\omega)^r I^{n-r} \operatorname{erfc}(\omega c_0) \sum_{\alpha_i} \frac{c_1^{\alpha_1} c_2^{\alpha_2} \ldots c_\mu^{\alpha_\mu}}{\alpha_1!\,\alpha_2!\ldots\alpha_\mu!} \tag{15}$$

in which $\mu_1 = \min(N-n, n)$ and

$$\alpha_1 + \alpha_2 + \ldots + \alpha_\mu = r; \qquad \alpha_1 + 2\alpha_2 + \ldots + \mu\alpha_\mu = N - n\,.$$

Thus, from (10), one has for $N = 0$,

$$a_0 = \frac{(1 - T_0)}{\operatorname{erf}(c_0)}\,; \qquad b_0 = \frac{1}{\operatorname{erfc}(\omega c_0)}\,, \tag{16}$$

and for $N \geqq 1$

$$\sum_{n=0}^{\infty} \{T_n A_{2n}^N(1) + a_n[A_n^N(1) - \tfrac{1}{2}(1 + (-1)^n)\, B_n^N(1)]\} = 0\,, \tag{17}$$

$$\sum_{n=0}^{\infty} b_n B_n^N(\omega) = 0\,. \tag{18}$$

Also from (12), we obtain for $N \geqq 1$

$$\sum_{n=0}^{\infty} \{T_n A_{2n+1}^N(1) + a_n[A_{n+1}^N(1) + \tfrac{1}{2}(1 - (-1)^n)\, B_n^N]\} = -\omega \sum_{n=0}^{\infty} b_n B_{n+1}^N(\omega)\,. \tag{19}$$

From this set of algebraic equations, (16–19), a_n, b_n and c_n can be determined sequentially.

Convergence of the series (8) and (11) can also be investigated in the manner of TAO [3]. However, we shall note only that τ_i, u_i and $s(t)$ are bounded at all times.

We note that for the case where $T(t) = \text{constant} = \tau_y T_0$ all the coefficients $a_n = b_n = c_n = 0$ für $n \geqq 1$. From (16) and set $N = 1$ in (19), we have the following equation for c_0

$$T_0 = \left(\sqrt{\pi}\,(1 - \omega^2)\, c_0 \operatorname{erf} c_0 + e^{-c_0^2} + \frac{\omega \operatorname{erf} c_0}{\operatorname{erfc}(\omega c_0)}\, e^{-\omega^2 c_0^2}\right) e^{c_0^2}, \tag{20}$$

This equation is identical in form to that of [2] where it was shown that, provided $T_0 > 1$, there is a solution to (20) for all ratio of η_1/η_2. In particular, for $c_0 = 1$, $T_0 = 5.27176$ if $\eta_1/\eta_2 = 0.01$ and $T_0 = 5.06022$ if $\eta_1/\eta_2 = 0$. If $T(t)$ is not a constant, by setting $N = 1, 2, \ldots$ in (17) and (18) and $N = 2, 3, \ldots$ in (19), the coefficients a_n, b_n, c_n, $n \geqq 1$ can be determined, e.g. $b_1 = b_0 B_0^1(\omega)/B_1^1(\omega)$, etc.

The case where a time-dependent velocity $U(t) = \sum_{n=0} U_n t^n/n!$ is prescribed at the boundary $y = 0$ can also be accommodated by TAO's [3] method. In this case it can be shown that $U_0 = 0$ for the existence of the solution. Thus, it is a special case of (13) and need not be treated separately.

References

1 GARTLING, D. K., Sandia National Laboratories, Albuquerque, New Mexico, Report, SANO 77–1333, 1977.
2 PHAN-THIEN, N., ASME J. Applied Mech., 50 (1983) 229.
3 TAO, L. N., Quart. Appl. Math. **36** (1978) 223; Z.A.M.P. **30** (1979) 416; ASME J. Appl. Mech. **48** (1981) 732.
4 BINGHAM, E. C., Fluidity and Plasticity, McGraw-Hill, New York (1922) 215.
5 OLDROYD, J. C., Proc. Camb. Phil. Soc. **43** (1947) 100.

Received 1983 January 5

Address: Dr. N. PHAN-THIEN, Mechanical Engineering Department, Sydney University, New South Wales 2006, Australia

Buchbesprechungen

Blanchard, P. / Brüning, E., Direkte Methoden der Variationsrechnung. Ein Lehrbuch. Wien-New York, Springer-Verlag 1982. IX, 279 S., 7 Abb., DM 98,–. S 686.––. US $ 40.90. ISBN 3-211-81692-5

Dieses Buch ist aus Vorlesungen für Mathematik- und Physikstudenten hervorgegangen, die über Grundkenntnisse der linearen Funktionalanalysis und der Differentialgeometrie verfügten. Sein Ziel ist es, „einen Überblick über die wichtigsten Problemstellungen und zentralen Ideen der sogenannten direkten Methoden der Variationsrechnung zu geben und einen möglichst elementaren und klaren Zugang dazu anzubieten". Den Autoren ist zu bescheinigen, daß sie dieses Ziel erreicht haben.

Das Buch beginnt mit einer schönen Einleitung zur Geschichte und den Zielen der Variationsrechnung. Der eigentliche Stoff ist in IX Kapitel gegliedert, die durch vier Anhänge mit benötigten Grundtatsachen über Banach-Räume, Stetigkeit und Halbstetigkeit, Kompaktheit in Banach-Räumen sowie über Sobolev-Räume ergänzt werden. Im ersten Kapitel wird der Weierstraßsche Fundamentalsatz der Variationsrechnung und seine Verallgemeinerung auf Banach-Räume behandelt. Kurz wird auf Anwendungen in der linearen Optimierung und auf das Ritzsche Approximationsverfahren eingegangen. Kapitel II bringt Differentialrechnung in Banach-Räumen: Fréchet-, Gateaux-Ableitung, Beziehungen zwischen der Konvexität differenzenzierbarer Funktionale und der Monotonie ihrer Gradienten. Im III. Kapitel werden Extrema differenzierbarer Funktionale analysiert und notwendige und hinreichende Bedingungen formuliert. Ein Kernstück des Buches ist das Kapitel IV, in dem Extremalprobleme unter Nebenbedingungen betrachtet werden. Die Methode der Lagrangeschen Multiplikatoren wird mittels des fundamentalen Satzes von LJUSTERNIK begründet. Das V. Kapitel ist klassischen Variationsproblemen gewidmet: Hamiltonsches Prinzip, Symmetrien und Erhaltungssätze in der klassischen Mechanik, das Problem der Brachstochrone, Feldtheorie, der Noethersche Satz in der klassischen Feldtheorie. Kapitel VI enthält die variationstheoretische Behandlung linearer Rand- und Eigenwertprobleme: Spektralsatz für kompakte selbstadjungierte Operatoren, Minimax-Prinzip, Satz von LAX-MILGRAM, Spektrum linearer elliptischer Differentialoperatoren, variationstheoretische Lösung linearer parabolischer Gleichungen u. a. In Kapitel VII wird die auf VISIK, BROWDER und MINTY zurückgehende Monotonietheorie nichtlinearer elliptischer Randwertprobleme dargestellt. Kapitel VIII enthält die Ljusternik-Schnirelman-Theorie kritischer Punkte und ihre Anwendung auf nichtlineare elliptische Eigenwertprobleme. Schließlich werden im IX. Kapitel in den letzten Jahren von LIEB, SIMON, BREZIS u. a. mit Variationsmethoden erzielte Ergebnisse zur Thomas-Fermi-Theorie quantenmechanischer Vielelektronensysteme entwickelt.

Aus dieser kurzen Inhaltsangabe ist ersichtlich, daß das vorliegende Buch in geschickter Gliederung sowohl die zentralen klassischen Resultate und Methoden als auch relevante moderne Entwicklungen in der Variationsrechnung präsentiert. Die Darstellung des Buches ist klar und korrekt. Das Verständnis der Beweise wird durch viele heuristische Erläuterungen im Text erleichtert und vertieft.

In den Anwendungen beschränkt sich das Buch bewußt auf die mathematische Physik. Dabei wird zudem auf Beispiele aus der Kontinuumsmechanik, einem der traditionellen und aktuellen Hauptanwendungsgebiete für Variationsmethoden, verzichtet. Dieser Verzicht hat aus der Sicht der Referenten die bedauerliche Konsequenz, daß im Rahmen der Thematik des Buches wichtigen Gebieten, wie etwa der Dualitätstheorie und numerischen Aspekten direkter Variationsmethoden, nicht die gebührende Beachtung geschenkt wird. Als symptomatisch sei erwähnt, daß einschlägige Standardwerke wie etwa EKELAND, TEMAM: Convex Analxsis and Variational Problems, North-Holl. Publ. Comp.-Amsterdam, Oxford (1976) und MICHLIN: Variationsmethoden der mathematischen Physik, Akad.-Verlag Berlin 1962 in den ansonsten sorgfältig ausgewählten Literaturhinweisen an den Enden der Kapitel fehlen. Ungeachtet dieser subjektiven Vorbehalte kann das vorliegende Buch als moderne Einführung in das schöne und aktuelle Gebiet der Variationsmethoden Lernenden und Lehrenden warm empfohlen werden.

Berlin H. GAJEWSKI

Brockett, R. W. / Millmann, R. S. / Sussmann, H. J. (eds.), Differential Geometric Control Theory. Boston-Basel-Stuttgart, Birkhäuser 1983. VII, 340 S., ISBN 3-7643-3091-0 (Progress in Mathematics 27)

Vom 28. 6. bis 2. 7. 1982 fand an der Technologischen Universität von Michigan/USA eine beachtenswerte Spezialtagung über Steuerungstheorie mit ca. 80 Teilnehmern statt. Den umfangreichsten Beitrag im Vortragsprogramm bildeten Übersichtsvorträge von H. J. SUSSMANN über "Lie Brackets, Real Analyticity and Geometric Control" R. B. GARDNER über „Differential Geometric Methods Interfacing Control Theorie". Die Vorträge von SUSSMANN bieten, kurz gesagt, eine Darstellung der Theorie steuerbarer Systeme auf C^∞-Mannigfaltigkeiten mit entsprechenden mannigfaltigkeitstheoretischen Versionen der Aussagen über erreichbare Mengen, (vollständig) steuerbare Systeme, Maximum-Prinzip, dynamische Programmierung, reguläre Synthese, Systemäquivalenz, Theorie der Realisierung usw.. Die Vorträge von GARDNER befassen sich mit äußerer Differentiation in der Variationsrechnung, Pfaffschen Systemen, Strahlenbündeln, Mongeschen Systemen in der Steuerungstheorie, der Cartanschen Methode der Äquivalenz usw. in der Sprache der heutigen Differentialgeometrie. Die übrigen Vorträge, die in dem Band Aufnahme fanden, lauten: BROCKETT, Asymptotic Stability and Feedback Stabilization; BYRNESS, Control Theory, Inverse Spectral Problems, and Real Algebraic Geometry; BYRNESS und KRENER, On the Existence of Globally (f, g)-Invariant Distributions; FLEISS, On the Relationship Between the Local Realization of Nonlinear Systems, Filtered, Transitive Lie Algebras; GILKEY, Secondary Characteristic Classes of Locally Flat Bundles Over Lens Spaces in Dimension 3; HARDT, Some Analytic Bounds for Subanalytic Sets; HUNT, SU und MEYER; Design for Multi-Input Nonlinear Systems; JURDJEVIC und SALLET, Controllability of Affine Systems; KUPKA, Generic Properties of Extremals in Optimal Control Problems; SU, MEYER und HUNT; Robustness in Nonlinear Control.

Die Vortragssammlung wird das weitere Eindringen mannigfaltigkeitstheoretischer Ideen in die Theorie der (optimalen) Steuerung als erfolgversprechende Hilfsmittel bei der Darstellung und Lösung von Problemen fördern und verdient die Aufmerksamkeit aller Fachleute.

Greifswald L. BITTNER

Cremers, A. B. / Kriegel, H. P. (eds.), Theoretical Computer Science. 6th GI-Conference, Dortmund 1983. Berlin-Heidelberg-New York, Springer-Verlag 1982. X, 367 S., DM 39,50. US $ 15.80. ISBN 3-540-11973-6 (Lecture Notes in Computer Science 145)

Die im Tagungsband enthaltenen Arbeiten gehören zu den Gebieten Kompliziertheitstheorie (10 Arbeiten, davon 4 Arbeiten über schnelle Algorithmen für konkrete Probleme und 6 Arbeiten über mehr theoretische Fragen der Kompliziertheitstheorie), Algorithmen- und Berechnungstheorie (5), Sprachtheorie (4), Datenstrukturen (4), Concurrency (4) und Sonstiges (4).

Über die Hälfte der Arbeiten (10) kommen aus der BRD; es folgen Frankreich (6) und USA (4). Als eingeladene Referenten sprachen R. L. CONSTABLE über partielle Funktionen in konstruktiven formalen Theorien, C. P. SCHNORR über einen Monte-Carlo-Faktorisierungs-Algorithmus mit endlichem Speicherbedarf, und C. H. PAPADIMITRIOU gab einen Überblick über die Theorie der Concurrency-Kontrolle.

Jena K. WAGNER

Morris, J. Ll., Computational Methods in Elementary Numerical Analysis. Chichester, John Wiley & Sons 1983. XII, 410 S., £ 9.95. ISBN 0-471-10420-5

Kaum der Text an sich, wohl aber die ganze Anlage dieses Lehrbuches der Methoden der numerischen Mathematik, zwingend jedenfalls die vielen Übungsaufgaben setzen die Kenntnis einer der gängigen Programmiersprachen und die Benutzung eines Computers voraus. Der Leser soll lernen, erfolgreich computergestützt numerisch zu arbeiten, wozu es ja nicht ausreicht, sich irgendwoher Software „verschreiben" zu lassen. Fertige Programme enthält das Buch dann auch folgerichtig nicht, der Titel könnte dahingehend mißverstanden werden. Ferner rangieren Interpretation und Motivation der vorgeführten Methoden vor Vollständigkeit der Beweissammlung. Schließlich meint „elementar" hier auch, daß ein realistisch angelegter Kurs notwendig Einiges weglassen muß, im vorliegenden Fall z. B. partielle Differentialgleichungen und Optimierung. Auch so reicht der Stoff für eine solide zweisemestrige Vorlesung mit Übungen.

Berlin W. ZEISS

Arató, M., Linear Stochastic Systems with Constant Coefficients. A Statistical Approach. Berlin-Heidelberg-New York, Springer-Verlag 1982. IX, 309 S., DM 38,50. US $ 15.40. ISBN 3-540-12090-4 (Lecture Notes in Control and Information Sciences 45)

Das vorliegende Buch stellt sich zum Ziel, die häufig existierende Kluft zwischen einer recht weit entwickelten statistischen Theorie stochastischer Prozesse und einem breiten Feld praktischer Anwendungsmöglichkeiten zu verringern. Diesem wichtigen Anliegen wird die hier vorliegende umfangreiche Arbeit voll gerecht und entspricht somit sowohl dem Anliegen „theoretischer" Statistiker, ihre Resultate angewendet zu sehen als auch dem Wunsch vieler Anwender, mit den neuesten Ergebnissen der mathematischen Theorie vertraut gemacht zu werden. (Allein die sich im Anhang befindende umfangreiche Bibliographie weiterführender Arbeiten zu Linearen Stochastischen Systemen mit konstanten Koeffizienten ist eine Fundgrube für jeden interessierten Leser und ein Zeichen dafür, wie gründlich der Autor bemüht war, moderne Resultate auf diesem Gebiet in sein Buch einzuarbeiten).

Dem Anliegen des Autors, einen möglichst breiten Leserkreis zu erreichen, angefangen vom Mathematikstudenten, über Statistiker, theoretische Mechaniker und Praktiker entspricht der Aufbau des Buches. In dessen erstem Kapitel wird eine Vielzahl von Beispielen gegeben, wie sich reale Prozesse durch Lineare Stochastische Prozesse mit konstanten Koeffizienten beschreiben lassen und welche statistischen Problemstellungen sich folgerichtig hieraus ergeben. So ist der Leser ausreichend motiviert und vorbereitet für das zweite Kapitel, in dem bekannte Charakterisierungssätze für elementare Gaußsche Prozesse bei kontinuierlicher und diskreter Beobachtungszeit als Lösungen von Stochastischen Differential- und Differenzengleichungen angegeben und bewiesen werden. Der Zusammenhang zur Spektraltheorie wird ebenfalls hergestellt. Da wiederum jeweils spezielle Anwendungsbeispiele eingearbeitet werden, geht dieses Kapitel über das bloße Zusammenstellen bekannter Sätze weit hinaus. Im letzten Teil des Buches (Kapitel 3 und 4) sind statistische

Resultate bei der Untersuchung linearer stochastischer Systeme zusammengetragen. (Konstruktion und Eigenschaften von Schätzungen, Schwierigkeiten bei der rechentechnischen Realisierung derartiger Schätzvorschriften, asymptotische Aussagen hierzu, Hinweise auf noch offene Probleme).

Einzelne Abschnitte des Buches werden zwar von bereits bekannten mathematischen Büchern wie beispielsweise von LIPZER/SCHIRJAJEV, ROZANOW, BALAKRISHNAN u. a. überdeckt, in seiner Komplexität und Vollständigkeit dürfte dieses Buch alle bisherigen Veröffentlichungen zum gleichen Thema übertreffen. Es ist daher jedem Wissenschaftler oder Praktiker, der mit entsprechenden stochastischen Systemen zu tun hat, wärmstens zu empfehlen.

Berlin B. BELLACH

Eringen, A. C., Mechanics of Continua. 2nd edition. Huntington, New York, Robert E. Krieger Publ. Co. 1980. 608 S., US $ 39.50. ISBN 0-88275-663-X

Der international bekannte Autor hat mit dieser zweiten, überarbeiteten Ausgabe der „Mechanik der Kontinua" erneut ein erstklassiges Fachbuch vorgelegt. Die ersten drei Kapitel sind den allgemeinen Grundlagen gewidmet, die notwendigerweise jedes in sich geschlossene Lehrbuch der Kontinuumsmechanik enthalten sollte. Die Darstellung des Stoffes ist übersichtlich und zweckmäßig. Am Ende dieser und auch der nachfolgenden Kapitel werden jeweils Aufgaben zur behandelten Problematik zusammengestellt. Darüber hinaus ist jedem Kapitel eine kurze verbale Inhaltsangabe vorangestellt. Kapitel 4 des Buches enthält eine schöne Übersicht über wichtige Beziehungen der Kontinuumsthermodynamik. Daran schließt sich logisch die Stoffgesetzproblematik (konstitutive Gleichungen) an. Klassischen Anwendungsgebieten der Kontinuumstheorie wie Elastizitätstheorie, Strömungsmechanik, Thermoelastizität und Viskoelastizität sind die Kapitel 6 bis 9 gewidmet. Auch hier versteht es der Autor in ausgezeichneter Weise, von den theoretischen Grundlagen bis hin zu einfachen Anwendungen verständliche und doch anspruchsvolle Mechanik darzubieten. Kapitel 10 ist der Elektrodynamik der Kontinua gewidmet, wobei hier die Kopplung zur Mechanik behandelt wird. Es ist nach Ansicht des Rezensenten auch angesichts der gegenwärtig zu beobachtenden stürmischen internationalen Entwicklung dieses Fachgebietes zwischen Mechanik, Thermodynamik und Elektrodynamik der Höhepunkt des vorliegenden Buches. Das Interesse an derartigen Fragestellungen der „Feldkopplungsproblematik" dürfte in den nächsten Jahren weiter zunehmen, so daß es gut ist, wenn sich auch Lehrbücher der Kontinuumsmechanik stärker mit diesen Fragen befassen. Das in erster Linie als Lehrbuch zu verstehende Werk ERINGENS ist jedoch auch für den Forscher ein wertvolles Hilfsmittel, wenn auch angesichts der Breite der Themen nicht immer bis zur aktuellen Forschung vorgedrungen werden kann. Ein Schönheitsfehler sind die nicht wenigen Druckfehler. Die vierseitige Druckfehlerliste repräsentiert im Gegensatz zum Inhalt des Buches sicher kein fortgeschrittenes internationales Niveau. Dennoch sollte das Buch einen breiten Leserkreis von Interessenten aus verschiedenen Gebieten der Naturwissenschaften, Mathematik und Technik finden.

Karl-Marx-Stadt B. MICHEL

Vukobratović, M. / Potkonjak, V., Scientific Fundamentals of Robotics 1 (in 5 Bänden): Dynamics of Manipulation Robots, Theory and Application. Berlin-Heidelberg-New York, Springer-Verlag 1982. XIII, 303 S., 149 Abb., DM 82,–. US $ 36.40. ISBN 3-540-11628-1 (Communications and Control Engineering Series)

Die vorliegende Monografie stellt den ersten Band einer Reihe „Wissenschaftliche Grundlagen der Robotertechnik" dar, mit der eine vorwiegend anwendungsorientierte Darstellung der Grundlagen der Dynamik und der Synthese von Roboterstrukturen geplant ist. Das Buch ist aus einer mehrjährigen Forschungstätigkeit eines Kollektivs des Mihailo-Pupin-Instituts unter der Leitung von Prof. M. VUKOBRATOVIĆ entstanden und gibt einen guten Überblick über moderne Entwicklungstendenzen auf dem Gebiet der Dynamik von aktiven räumlichen Mechanismen, der Synthese ihrer Bewegung sowie des rechnerunterstützten Entwurfs.

Probleme des Steuerentwurfs auf der Basis dynamischer Modellgleichungen sind gekennzeichnet durch die variable Struktur des Roboters, durch starke Nichtlinearitäten in den Gleichungen sowie durch zusätzliche Besonderheiten, die aus der Redundanz folgen. Hauptanliegen des Buches ist die Darstellung computerorientierter Methoden, angefangen von der automatischen Aufstellung der Bewegungsgleichungen bis zum Computer-Aided-Design.

Den zentralen Platz nimmt das Kap. 2 ein, in dem die autom. Generisierung und Lösung der Bewegungsgleichungen behandelt wird. Es ist das Verdienst der Autoren, erstmalig eine zusammenhängende Darstellung der Möglichkeiten — ausgehend von den Grundprinzipien der Mechanik — gegeben zu haben. Die Methoden sind in 3 Gruppen eingeteilt: 1. Methoden auf der Grundlage des d'Alembertschen Prinzips und der Newton-Euler-Gleichung. 2. Methode auf der Grundlage der Lagrangeschen Gleichungen 2. Art. 3. Methoden auf der Grundlage des Gaußschen Prinzips und der Gibbs-Appel-Gleichungen. Hervorhebenswert sind vor allem die Ausführungen zur Anwendung des Gaußschen Prinzips und der Gibbs-Appelschen Gleichungen für die Generierung der Modellgleichungen.

Kapitel 3 enthält Anwendungsbeispiele für die entwickelten Methoden, insbesondere zur Synthese der funktionell bedingten Bewegungen der verschiedenen Roboterkonfigurationen. Es beinhaltet Probleme der inversen Koordinatentransformation, die Berechnung der Antriebskräfte und -momente, Aussagen zum vollständigen dynamischen Modell (Manipulator einschließlich Antrieb) sowie zur Optimierung der nominellen Trajektorien. Kapitel 4 ist der dynamischen Modellbildung von Robotern mit elastischen Segmenten gewidmet. Daraus resultierende Steuerungs- und Regelungsanforderungen werden nicht behandelt, sondern Fragen des rechnerunterstützten Entwurfs (z. B. Ermittlung oberer Schranken für die Deformationen). Kapitel 5 enthält Aussagen zum Entwurf optimaler Trajektorien auf der Grundlage dynamischer Zielfunktionale und unter Beachtung robotertypischer Restriktionen.

Das Buch behandelt in klarer und übersichtlicher Weise einen aktuellen Forschungsgegenstand auf dem Gebiet der Roboterdynamik. Es kann allen an den wissenschaftlichen Grundlagen der Roboterdynamik Interessierten (Studenten im Spezialseminar, wissenschaftlich tätigen Ingenieuren) empfohlen werden.

Berlin B. HEIMANN

In der DDR beziehbar im Rahmen des Kontingents.

Zeitschrift für Angewandte Mathematik und Mechanik
Volume 64 1984 Number 3

Contents

Zeitschrift für Angewandte Mathematik und Mechanik is indexed in Current Contents/Physical, Chemical & Earth Sciences, and Science Citation Index.

FSC
www.fsc.org
MIX
Papier | Fördert gute Waldnutzung
FSC® C083411

Zeitfracht Medien GmbH
Ferdinand-Jühlke-Straße 7
99095 Erfurt, Deutschland
produktsicherheit@kolibri360.de